John Ruskin

The harbours of England

John Ruskin

The harbours of England

ISBN/EAN: 9783744737104

Printed in Europe, USA, Canada, Australia, Japan

Cover: Foto ©ninafisch / pixelio.de

More available books at **www.hansebooks.com**

THE

HARBOURS OF ENGLAND

THE
HARBOURS OF ENGLAND.

BY

JOHN RUSKIN,

HONORARY STUDENT OF CHRIST CHURCH, AND HONORARY FELLOW
OF CORPUS CHRISTI COLLEGE, OXFORD.

WITH

THIRTEEN ILLUSTRATIONS BY

J. M. W. TURNER, R.A.

EDITED BY

THOMAS J. WISE,

EDITOR OF

" A COMPLETE BIBLIOGRAPHY OF THE WRITINGS OF JOHN RUSKIN,"
ETC. ETC.

GEORGE ALLEN, SUNNYSIDE, ORPINGTON,

AND

156, CHARING CROSS ROAD, LONDON

1895

LIST OF ILLUSTRATIONS

EDITOR'S PREFACE

" TURNER'S *Harbours of England*," as it is generally called, is a book which, for various reasons, has never received from readers of Mr. Ruskin's writings the attention it deserves. True, it has always been sought after by connoisseurs, and collectors never fail with their eleven or twelve guineas whenever a set of Artist's Proofs of the First Edition of 1856 comes into the market. But to the General Reader the book with its twelve exquisitely delicate mezzotints—four of which Mr. Ruskin has declared to be among the very finest executed by Turner from his marine subjects—is practically unknown.

The primary reason for this neglect is not far to seek. Since 1877 no new edition of the work has been published, and thus it has gradually passed from public knowledge, though still regarded with lively interest by those to whom Mr. Ruskin's words—particularly words written in further unfolding of the subtleties of Turner's art—at all times appeal so strongly.

In his own preface Mr. Ruskin has told us all that in 1856 it was necessary to know of the genesis of the *Harbours*. That account may now be supplemented with the following additional facts. In 1826 Turner (in conjunction with Lupton, the engraver) projected and commenced a serial publication entitled *The Ports of England*. But both artist and engraver lacked the opportunity required to carry the undertaking to a successful conclusion, and three numbers only were completed. Each of these contained two engravings. Part I., introducing *Scarborough* and *Whitby*, duly appeared in 1826; Part II., with *Dover* and *Ramsgate*, in 1827; and in 1828 Part III., containing *Sheerness* and *Portsmouth*, closed the series.*

* To ornament the covers of these parts, Turner designed a vignette, which was printed upon the centre of the front wrapper of each. As *The Ports of England* is an exceptionally scarce book, and as the vignette can be obtained in no other form, a facsimile of it is here given. The original drawing was presented by Mr. Ruskin to the Fitz-William Museum, at Cambridge, where it may now be seen.

1826

To face p. x.

Twenty-eight years afterwards (that is, in 1856, five years after Turner's death) these six plates, together with six new ones, were published by Messrs. E. Gambart & Co., at whose invitation Mr. Ruskin consented to write the essay on Turner's marine painting which accompanied them. The book, a handsome folio, appears to have been immediately successful, for in the following year a second edition was called for. This was a precise reprint of the 1856 edition; but, unhappily, the delicate plates already began to exhibit signs of wear. The copyright (which had not been retained by Mr. Ruskin, but remained the property of Messrs. E. Gambart & Co.) then passed to Messrs. Day & Son, who, after producing the third edition of 1859, in turn disposed of it to Mr. T. J. Allman. Allman issued a fourth edition in 1872, and then parted with his rights to Messrs. Smith, Elder & Co., who in 1877 brought out the fifth, and, until now, last edition. Since that date the work has been out of print, and has remained practically inaccessible to the ordinary reader.

It is matter for congratulation that at length means have been found to bring *The Harbours of England* once more into currency, and to issue the book through Mr. George Allen at a price which will place it within the reach of the reading public at large.

The last edition of 1877, with its worn and "retouched" plates,* was published at twenty-five shillings; less than a third of that sum will suffice to procure a copy of this new issue in which the prints (save for their reduced size) more nearly approach the clearness and beauty of the originals of 1856 than any of the three editions which have immediately preceded it.

I have before me the following interesting

---

* By this time (1877) the plates had become considerably worn, and were accordingly "retouched" by Mr. Chas. A. Tomkins. But such retouching proved worse than useless. The delicacy of the finer work had entirely vanished, and the plates remained but a ghost of their former selves, such as no one would recognise as doing justice to Turner. The fifth is unquestionably the least satisfactory of the five original editions containing Lupton's engravings.

letter addressed by Mr. Ruskin's father to Mr. W. Smith Williams, for many years literary adviser to Messrs. Smith, Elder & Co. :—

" CHAMOUNI, *August 4th*, 1856.

"MY DEAR SIR,— I hear that in *The Athenæum* of 26th July there is a good article on my son's *Harbours of England*, and I should be greatly obliged by Mr. Gordon Smith sending me that number. . . .

"The history of this book, I believe, I told you. Gambart, the French publisher and picture dealer, said some 18 months ago that he was going to put out 12 Turner plates, never published, of English Harbours, and he would give my son two good Turner drawings for a few pages of text to illustrate them.* John agreed, and wrote the text, when poorly in the spring of 1855, at Tunbridge Wells; and it seems the work has just come out. It was in my opinion an extremely well done

* Mr. E. Gambart (who is still living) states that, to the best of his recollection, he paid Mr. Ruskin 150 guineas for his work. Probably this was the price originally agreed upon, the two Turner drawings being ultimately accepted as a more welcome and appropriate form of remuneration.

thing, and more likely, as far as it went,
if not to be extremely popular, at least
to be received without cavil than anything
he had written. If there is a very favour-
able review in *The Athenæum* . . . it may
tend to disarm the critics, and partly influence
opinion of his larger works. . . .—With our
united kind regards,

      "Yours very truly,
          "JOHN JAMES RUSKIN."

In all save one particular the Text here
given follows precisely that of the previous
issues. It has been the good fortune of the
present Editor to be able to restore a characteristic passage suppressed from motives
of prudence when the work was originally
planned.* The proof-sheets of the first
edition, worked upon by Mr. Ruskin, were
given by him to his old nurse Anne.† She,
fortunately, carefully preserved them, and in
turn gave them to Mr. Allen, some ten years
before he became Mr. Ruskin's publisher.

---

* See *post*, p. 33.
† See *Præterita*. She died March 30th, 1871.

Respecting this lower kind of ship-painting, it is always matter of wonder to me that it satisfies sailors. Some years ago I happened to stand longer than pleased my pensioner guide before Turner's "Battle of Trafalgar," at Greenwich Hospital; a picture which, at a moderate estimate, is simply worth all the rest of the hospital, ground, walls, pictures, and models put together. My guide supposing me to be detained by indignant wonder at seeing it in so good a place, assented to my supposed sentiments

These proofs had been submitted as they came from the press to Mr. W. H. Harrison (well known to readers of *On the Old Road*, &c., as " My First Editor "), who marked them freely with notes and suggestions. To one passage he appears to have taken so decided an objection that its author was prevailed upon to delete it. But, whilst deferring thus to the judgment of others, and consenting to remove a sentence which he doubtless regarded with particular satisfaction as expressing a decided opinion upon a favourite picture, Mr. Ruskin indulged in one of those pleasantries which now and again we observe in his informal letters, though seldom, if ever, in his serious writings. In the margin, below the cancelled passage, he wrote boldly: " *Sacrificed to the Muse of Prudence. J. R.*" *

That Mr. Harrison was justified in raising objection to this " moderate estimate " of Turner's picture will, I think, be readily allowed. In those days Mr. Ruskin's influence was,

* The accompanying illustration is a facsimile of the portion of the proof sheet described above—slightly reduced to fit the smaller page.

comparatively speaking, small; and the expression of an opinion which heaped praise
upon the single painting of a partially understood painter at the expense of a great and
popular institution would only have served
to arouse opposition, and possibly to attract
ridicule. It is different to-day. We know
the keen enthusiasm of the author of *The
Seven Lamps*, and have seen again and again
how he expresses himself in terms of somewhat exaggerated admiration when writing of
a painter whom he appreciates, or a picture
that he loves. To us this enthusiasm is an
attractive characteristic. It has never been
permitted to distort the vision or cloud the
critical faculty; and we follow the teaching
of the Master all the more closely because
we feel his fervour, and know how completely
he becomes possessed with a subject which
appeals to his imagination or his heart. I
have therefore not scrupled to revive the
words which he consented to immolate at the
shrine of Prudence.

It is not my province here to enter into
any criticism of the pages which follow;

but, for the benefit of those who are not versed in the minutiæ of Shelleyan topics, a word may be said regarding Mr. Ruskin's reference * to the poet who met his death in the Bay of Spezzia. The *Don Juan* was no "traitorous" craft. Fuller and more authentic information is to hand now than the meagre facts at the disposal of a writer in 1856; and we know that the greed of man, and not the lack of sea - worthiness in his tiny vessel, caused Percy Shelley to

> ". . . Suffer a sea change
> Into something rich and strange."

There is, unhappily, no longer any room for doubt that the *Don Juan* was wilfully run down by a felucca whose crew coveted the considerable sum of money they believed Byron to have placed on board, and cared nothing for the sacrifice of human life in their eagerness to seize the gold.

The twelve engravings, to which reference has already been made, have been reproduced

* See *post*, p. 6.

b

by the photogravure process from a selected
set of early examples ; and, in addition, the
plates so prepared have been carefully worked
upon by Mr. Allen himself.   It will thus be
apparent that everything possible has been
done to produce a worthy edition of a worthy
book, and to place in the hands of the public
what to the present generation of readers is
tantamount to a new work from a pen which—
alas !—has now for so long a time been still.

THOMAS J. WISE.

AUTHOR'S
ORIGINAL PREFACE

AUTHOR'S
ORIGINAL PREFACE

AMONG the many peculiarities which distinguished the late J. M. W. Turner from other landscape painters, not the least notable, in my apprehension, were his earnest desire to arrange his works in connected groups, and his evident intention, with respect to each drawing, that it should be considered as expressing part of a continuous system of thought. The practical result of this feeling was that he commenced many series of drawings,—and, if any accident interfered with the continuation of the work, hastily concluded them,—under titles representing rather the relation which the executed designs bore to the materials accumulated in his own mind, than the position which they could justifiably claim when

contemplated by others. The *River Scenery*
was closed without a single drawing of a
rapidly running stream; and the prints of his
annual tours were assembled, under the title
of the *Rivers of France*, without including
a single illustration either of the Rhone or the
Garonne.

The title under which the following plates
are now presented to the public, is retained
merely out of respect to this habit of Turner's.
Under that title he commenced the publication,
and executed the vignette for its title-page,
intending doubtless to make it worthy of
taking rank with, if not far above, the consistent and extensive series of the *Southern
Coast*, executed in his earlier years. But
procrastination and accident equally interfered
with his purpose. The excellent engraver
Mr. Lupton, in co-operation with whom the
work was undertaken, was unfortunately also
a man of genius, and seems to have been just
as capricious as Turner himself in the applica-
tion of his powers to the matter in hand. Had

one of the parties in the arrangement been a mere plodding man of business, the work would have proceeded; but between the two men of talent it came very naturally to a stand. They petted each other by reciprocal indulgence of delay; and at Turner's death, the series, so magnificently announced under the title of the *Harbours of England*, consisted only of twelve plates, all the less worthy of their high-sounding title in that, while they included illustrations of some of the least important of the watering-places, they did not include any illustration whatever of such harbours of England as Liverpool, Shields, Yarmouth, or Bristol. Such as they were, however, I was requested to undertake their illustration. As the offer was made at a moment when much nonsense, in various forms, was being written about Turner and his works; and among the twelve plates there were four * which I considered among the very finest that had been executed from his marine subjects, I accepted the trust; partly to prevent the really valuable

* Portsmouth, Sheerness, Scarborough, and Whitby.

series of engravings from being treated with injustice, and partly because there were several features in them by which I could render more intelligible some remarks I wished to make on Turner's marine painting in general.

These remarks, therefore, I have thrown together, in a connected form ; less with a view to the illustration of these particular plates, than of the general system of ship-painting which was characteristic of the great artist.   I have afterwards separately noted the points which seemed to me most deserving of attention in the plates themselves.

Of archæological information the reader will find none.   The designs themselves are, in most instances, little more than spirited sea-pieces, with such indistinct suggestion of local features in the distance as may justify the name given to the subject; but even when, as in the case of the Dover and Portsmouth, there is something approaching topographical detail, I have not considered it necessary to lead the reader into inquiries which certainly Turner himself never thought of; nor do I suppose it would materially add to the interest of these cloudy distances or rolling seas, if I had the time—which I have not—to collect the most complete information respecting the raising of Prospect Rows, and the establishment of circulating libraries.

DENMARK HILL.
[1856.]

c

THE

HARBOURS OF ENGLAND

THE

HARBOURS OF ENGLAND

OF all things, living or lifeless, upon this
strange earth, there is but one which, having
reached the mid-term of appointed human en-
durance on it, I still regard with unmitigated
amazement. I know, indeed, that all around
me is wonderful—but I cannot answer it with
wonder :—a dark veil, with the foolish words,
NATURE OF THINGS, upon it, casts its deaden-
ing folds between me and their dazzling
strangeness. Flowers open, and stars rise,
and it seems to me they could have done no
less. The mystery of distant mountain-blue
only makes me reflect that the earth is of
necessity mountainous ;—the sea-wave breaks
at my feet, and I do not see how it should have
remained unbroken. But one object there is
still, which I never pass without the renewed

A

wonder of childhood, and that is the bow of a Boat. Not of a racing-wherry, or revenue cutter, or clipper yacht; but the blunt head of a common, bluff, undecked sea-boat, lying aside in its furrow of beach sand. The sum of Navigation is in that. You may magnify it or decorate as you will: you do not add to the wonder of it. Lengthen it into hatchet-like edge of iron,—strengthen it with complex tracery of ribs of oak,—carve it and gild it till a column of light moves beneath it on the sea, —you have made no more of it than it was at first. That rude simplicity of bent plank, that can breast its way through the death that is in the deep sea, has in it the soul of shipping. Beyond this, we may have more work, more men, more money; we cannot have more miracle.

For there is, first, an infinite strangeness in the perfection of the thing, as work of human hands. I know nothing else that man does, which is perfect, but that. All his other doings have some sign of weakness, affectation, or ignorance in them. They are overfinished or underfinished; they do not quite answer their end, or they show a mean vanity in answering it too well.

But the boat's bow is naïvely perfect: complete without an effort. The man who made it knew not he was making anything beautiful, as he bent its planks into those mysterious, ever-changing curves. It grows under his hand into the image of a sea-shell; the seal, as it were, of the flowing of the great tides and streams of ocean stamped on its delicate rounding. He leaves it when all is done, without a boast. It is simple work, but it will keep out water. And every plank thenceforward is a Fate, and has men's lives wreathed in the knots of it, as the cloth-yard shaft had their deaths in its plumes.

Then, also, it is wonderful on account of the greatness of the thing accomplished. No other work of human hands ever gained so much. Steam-engines and telegraphs indeed help us to fetch, and carry, and talk; they lift weights for us, and bring messages, with less trouble than would have been needed otherwise; this saving of trouble, however, does not constitute a new faculty, it only enhances the powers we already possess. But in that bow of the boat is the gift of another world. Without it, what prison wall would be so strong as that "white and wailing fringe" of sea. What maimed creatures were we all, chained to our rocks, Andromeda-like, or wandering by the endless shores, wasting our incommunicable strength, and pining in hopeless watch of unconquerable waves ? The nails that fasten together the planks of the boat's bow are the rivets of the fellowship of the world. Their iron does more than draw lightning out of heaven, it leads love round the earth.

Then also, it is wonderful on account of the greatness of the enemy that it does battle with. To lift dead weight; to overcome length of languid space; to multiply or systematise a given force; this we may see done by the bar, or beam, or wheel, without wonder. But to war with that living fury of waters, to bare its breast, moment after moment, against the unwearied enmity of ocean,—the subtle, fitful, implacable smiting of the black waves, provoking each other on, endlessly, all the infinite march of the Atlantic rolling on behind them to their help,—and still to strike them back into a wreath of smoke and futile foam, and win its way against them, and keep its charge of life from them ;—does any other soulless thing do as much as this ?

I should not have talked of this feeling of mine about a boat, if I had thought it was mine only ; but I believe it to be common to all of us who are not seamen. With the seaman, wonder changes into fellowship and close affection ; but to all landsmen, from youth upwards, the boat remains a piece of enchantment ; at least unless we entangle our vanity in it, and refine it away into mere lath, giving up all its protective nobleness for pace. With those in whose eyes the perfection of a boat is swift fragility, I have no sympathy. The glory of a boat is, first its steadiness of poise—its assured standing on the clear softness of the abyss ; and, after that, so much capacity of progress by oar or sail as shall be consistent with this defiance of the treachery of the sea. And, this being understood, it is very notable how commonly the poets, creating for themselves an ideal of motion, fasten upon the charm of a boat. They do not usually express any desire for wings, or, if they do, it is only in some vague and half-unintended phrase, such as " flit or soar," involving wingedness. Seriously, they are evidently content to let the wings belong to Horse, or Muse, or Angel, rather than to themselves; but they all, somehow or other, express an honest wish for a Spiritual Boat. I will not dwell on poor Shelley's paper navies, and seas of quicksilver, lest we should begin to think evil of boats in general because of that traitorous one in Spezzia Bay; but it is a triumph to find the pastorally minded Wordsworth imagine no other way of visiting the stars than in a boat " no bigger than the crescent moon";* and to find Tennyson—although his boating, in an ordinary way, has a very marshy and punt-like character—at last, in his highest inspiration, enter in where the wind began "to sweep a music out of sheet and shroud." † But the chief triumph of all is in Dante. He had known all manner of travelling; had been borne through vacancy on the shoulders of chimeras, and lifted through upper heaven in the grasp of its spirits; but yet I do not remember that he ever expresses any positive *wish* on such matters, except for a boat.

* Prologue to *Peter Bell*.   † *In Memoriam*, ci.

' Guido, I wish that Lapo, thou, and I,
    Led by some strong enchantment, might ascend
    A magic ship, whose charmëd sails should fly
    With winds at will where'er our thoughts might
      wend,
So that no change nor any evil chance
    Should mar our joyous voyage ; but it might be
    That even satiety should still enhance
    Between our souls their strict community :
And that the bounteous wizard then would place
    Vanna and Bice, and our Lapo's love,
    Companions of our wandering, and would grace
With passionate talk, wherever we might rove,
Our time, and each were as content and free
As I believe that thou and I should be."

And of all the descriptions of motion in the *Divina Commedia*, I do not think there is another quite so fine as that in which Dante has glorified the old fable of Charon by giving a boat also to the bright sea which surrounds the mountain of Purgatory, bearing the redeemed souls to their place of trial; only an angel is now the pilot, and there is no stroke of labouring oar, for his wings are the sails.

     " My preceptor silent yet
Stood, while the brightness that we first discerned
Opened the form of wings : then, when he knew
The pilot, cried aloud, ' Down, down ; bend low

Thy knees ; behold God's angel : fold thy hands :
Now shalt thou see true ministers indeed.
Lo ! how all human means he sets at nought ;
So that nor oar he needs, nor other sail
Except his wings, between such distant shores.
Lo ! how straight up to heaven he holds them reared,
Winnowing the air with those eternal plumes,
That not like mortal hairs fall off or change.'

"As more and more toward us came, more bright
Appeared the bird of God, nor could the eye
Endure his splendour near : I mine bent down.
He drove ashore in a small bark so swift
And light, that in its course no wave it drank.
The heavenly steersman at the prow was seen,
Visibly written blessed in his looks.
Within, a hundred spirits and more there sat."

I have given this passage at length, because
it seems to me that Dante's most inventive
adaptation of the fable of Charon to Heaven
has not been regarded with the interest that
it really deserves; and because, also, it is a
description that should be remembered by
every traveller when first he sees the white
fork of the felucca sail shining on the Southern
Sea. Not that Dante had ever seen such
sails ; * his thought was utterly irrespective of

---

* I am not quite sure of this, not having studied with any
care the forms of mediæval shipping ; but in all the MSS. I
have examined the sails of the shipping represented are square.

the form of canvas in any ship of the period;
but it is well to be able to attach this happy
image to those felucca sails, as they now float
white and soft above the blue glowing of the
bays of Adria. Nor are other images wanting
in them. Seen far away on the horizon, the
Neapolitan felucca has all the aspect of some
strange bird stooping out of the air and just
striking the water with its claws; while the
Venetian, when its painted sails are at full
swell in sunshine, is as beautiful as a butterfly
with its wings half-closed.* There is something also in them that might remind us of the
variegated and spotted angel wings of Orcagna,
only the Venetian sail never looks majestic;
it is too quaint and strange, yet with no
peacock's pride or vulgar gaiety,—nothing of
Milton's Dalilah :

> " So bedecked, ornate and gay
> Like a stately ship

* It is not a little strange that in all the innumerable paintings of Venice, old and modern, no notice whatever had been
taken of these sails, though they are *exactly* the most striking
features of the marine scenery around the city, until Turner
fastened upon them, painting one important picture, "The
Sun of Venice," entirely in their illustration.

Of Tarsus, bound for the Isles
Of Javan or Gadire
With all her bravery on and tackle trim,
Sails filled and streamers waving."

That description could only have been written
in a time of vulgar women and vulgar vessels.
The utmost vanity of dress in a woman of
the fourteenth century would have given no
image of "sails filled or streamers waving";
nor does the look or action of a really "stately"
ship ever suggest any image of the motion of
a weak or vain woman. The beauties of the
Court of Charles II., and the gilded galleys
of the Thames, might fitly be compared; but
the pomp of the Venetian fisher-boat is like
neither. The sail seems dyed in its fulness
by the sunshine, as the rainbow dyes a cloud;
the rich stains upon it fade and reappear, as
its folds swell or fall; worn with the Adrian
storms, its rough woof has a kind of noble
dimness upon it, and its colours seem as
grave, inherent, and free from vanity as the
spots of the leopard, or veins of the seashell.

Yet, in speaking of poets' love of boats, I
ought to have limited the love to *modern*
poets; Dante, in this respect, as in nearly

every other, being far in advance of his age. It is not often that I congratulate myself upon the days in which I happen to live; but I do so in this respect, that, compared with every other period of the world, this nineteenth century (or rather, the period between 1750 and 1850) may not improperly be called the Age of Boats; while the classic and chivalric times, in which boats were partly dreaded, partly despised, may respectively be characterised, with regard to their means of locomotion, as the Age of Chariots, and the Age of Horses.

For, whatever perfection and costliness there may be in the present decorations, harnessing, and horsing of any English or Parisian wheel equipage, I apprehend that we can from none of them form any high ideal of wheel conveyance; and that unless we had seen an Egyptian king bending his bow with his horses at the gallop, or a Greek knight leaning with his poised lance over the shoulder of his charioteer, we have no right to consider ourselves as thoroughly knowing what the word "chariot," in its noblest acceptation, means.

So, also, though much chivalry is yet left in us, and we English still know several things about horses, I believe that if we had seen Charlemagne and Roland ride out hunting from Aix, or Cœur de Lion trot into camp on a sunny evening at Ascalon, or a Florentine lady canter down the Val d'Arno in Dante's time, with her hawk on her wrist, we should have had some other ideas even about horses than the best we can have now. But most assuredly, nothing that ever swung at the quay sides of Carthage, or glowed with crusaders' shields above the bays of Syria, could give to any contemporary human creature such an idea of the meaning of the word Boat, as may be now gained by any mortal happy enough to behold as much as a New-castle collier beating against the wind. In the classical period, indeed, there was some importance given to shipping as the means of locking a battle-field together on the waves; but in the chivalric period, the whole mind of man is withdrawn from the sea, regarding it merely as a treacherous impediment, over which it was necessary sometimes to find conveyance, but from which the thoughts

were always turned impatiently, fixing themselves in green fields, and pleasures that may
be enjoyed by land—the very supremacy of
the horse necessitating the scorn of the sea,
which would not be trodden by hoofs.

It is very interesting to note how repugnant
every oceanic idea appears to be to the whole
nature of our principal English mediæval poet,
Chaucer.   Read first the Man of Lawe's Tale,
in which the Lady Constance is continually
floated up and down the Mediterranean, and
the German Ocean, in a ship by herself;
carried from Syria all the way to Northum-
berland, and there wrecked upon the coast;
thence yet again driven up and down among
the waves for five years, she and her child;
and yet, all this while, Chaucer does not let fall
a single word descriptive of the sea, or express
any emotion whatever about it, or about the
ship.   He simply tells us the lady sailed
here and was wrecked there; but neither
he nor his audience appear to be capable of
receiving any sensation, but one of simple
aversion, from waves, ships, or sands.   Compare with his absolutely apathetic recital, the
description by a modern poet of the sailing

of a vessel, charged with the fate of another
Constance :

> " It curled not Tweed alone, that breeze—
> For far upon Northumbrian seas
>   It freshly blew, and strong ;
> Where from high Whitby's cloistered pile,
> Bound to St. Cuthbert's holy isle,
>   It bore a bark along.
> Upon the gale she stooped her side,
> And bounded o'er the swelling tide
>   As she were dancing home.
> The merry seamen laughed to see
> Their gallant ship so lustily
>   Furrow the green sea foam."

Now just as Scott enjoys this sea breeze,
so does Chaucer the soft air of the woods ;
the moment the older poet lands, he is himself
again, his poverty of language in speaking of
the ship is not because he despises description,
but because he has nothing to describe.   Hear
him upon the ground in Spring :

> " These woodes else recoveren greene,
> That drie in winter ben to sene,
> And the erth waxeth proud withall,
> For sweet dewes that on it fall,
> And the poore estate forget,
> In which that winter had it set :
> And than becomes the ground so proude,
> That it wol have a newe shroude,

> And maketh so queint his robe and faire,
> That it had hewes an hundred paire,
> Of grasse and floures, of Inde and Pers,
> And many hewes full divers :
> That is the robe I mean ywis,
> Through which the ground to praisen is."

In like manner, wherever throughout his poems we find Chaucer enthusiastic, it is on a sunny day in the "good greenwood," but the slightest approach to the sea-shore makes him shiver; and his antipathy finds at last positive expression, and becomes the principal foundation of the Frankeleine's Tale, in which a lady, waiting for her husband's return in a castle by the sea, behaves and expresses herself as follows :—

> " Another time wold she sit and thinke,
> And cast her eyen dounward fro the brinke ;
> But whan she saw the grisly rockes blake,
> For veray fere so wold hire herte quake
> That on hire feet she might hire not sustene
> Than wold she sit adoun upon the grene,
> And pitously into the see behold,
> And say right thus, with careful sighes cold.
>   ' Eterne God, that thurgh thy purveance
> Ledest this world by certain governance,
> In idel, as men sain, ye nothing make.
> *But, lord, thise grisly fendly rockes blake,*

That semen rather a foule confusion
Of werk, than any faire creation
Of swiche a parfit wise God and stable,
Why han ye wrought this werk unresonable?'"

The desire to have the rocks out of her way
is indeed severely punished in the sequel of
the tale; but it is not the less characteristic of
the age, and well worth meditating upon, in
comparison with the feelings of an unsophisti-
cated modern French or English girl among
the black rocks of Dieppe or Ramsgate.

On the other hand, much might be said
about that peculiar love of *green fields and
birds* in the Middle Ages; and of all with
which it is connected, purity and health in
manners and heart, as opposed to the too
frequent condition of the modern mind—

"As for the birds in the thicket,
Thrush or ousel in leafy niche,
Linnet or finch—she was far too rich
To care for a morning concert to which
She was welcome, without a ticket." *

But this would lead us far afield, and the
main fact I have to point out to the reader is
the transition of human grace and strength

* Thomas Hood.

from the exercises of the land to those of the sea in the course of the last three centuries.

Down to Elizabeth's time chivalry lasted; and grace of dress and mien, and all else that was connected with chivalry. Then came the ages which, when they have taken their due place in the depths of the past, will be, by a wise and clear-sighted futurity, perhaps well comprehended under a common name, as the ages of Starch; periods of general stiffening and bluish-whitening, with a prevailing washer-woman's taste in everything; involving a change of steel armour into cambric; of natural hair into peruke; of natural walking into that which will disarrange no wristbands; of plain language into quips and embroideries; and of human life in general, from a green race-course, where to be defeated was at worst only to fall behind and recover breath, into a slippery pole, to be climbed with toil and contortion, and in clinging to which, each man's foot is on his neighbour's head.

But, meanwhile, the marine deities were incorruptible. It was not possible to starch the sea; and precisely as the stiffness fastened upon men, it vanished from ships. What had

B

once been a mere raft, with rows of formal
benches, pushed along by laborious flap of
oars, and with infinite fluttering of flags and
swelling of poops above, gradually began to
lean more heavily into the deep water, to
sustain a gloomy weight of guns, to draw back
its spider-like feebleness of limb, and open its
bosom to the wind, and finally darkened down,
from all its painted vanities into the long, low
hull, familiar with the overflying foam; that
has no other pride but in its daily duty and
victory; while, through all these changes, it
gained continually in grace, strength, audacity,
and beauty, until at last it has reached such a
pitch of all these, that there is not, except the
very loveliest creatures of the living world,
anything in nature so absolutely notable,
bewitching, and, according to its means and
measure, heart-occupying, as a well-handled
ship under sail in a stormy day.   Any ship,
from lowest to proudest, has due place in that
architecture of the sea; beautiful, not so much
in this or that piece of it, as in the unity
of all, from cottage to cathedral, into their
great buoyant dynasty.   Yet, among them, the
fisher-boat, corresponding to the cottage on

the land (only far more sublime than a cottage
ever can be), is on the whole the thing most
venerable. I doubt if ever academic grove
were half so fit for profitable meditation as the
little strip of shingle between two black, steep,
overhanging sides of stranded fishing-boats.
The clear, heavy water-edge of ocean rising
and falling close to their bows, in that un-
accountable way which the sea has always in
calm weather, turning the pebbles over and
over as if with a rake, to look for something,
and then stopping a moment down at the
bottom of the bank, and coming up again with
a little run and clash, throwing a foot's depth
of salt crystal in an instant between you and
the round stone you were going to take in
your hand; sighing, all the while, as if it
would infinitely rather be doing something
else. And the dark flanks of the fishing-
boats all aslope above, in their shining quietness, hot in the morning sun, rusty and
seamed with square patches of plank nailed
over their rents; just rough enough to let the
little flat-footed fisher-children haul or twist
themselves up to the gunwales, and drop
back again along some stray rope; just round

enough to remind us, in their broad and
gradual curves, of the sweep of the green
surges they know so well, and of the hours
when those old sides of seared timber, all
ashine with the sea, plunge and dip into the
deep green purity of the mounded waves more
joyfully than a deer lies down among the
grass of spring, the soft white cloud of foam
opening momentarily at the bows, and fading
or flying high into the breeze where the seagulls toss and shriek,—the joy and beauty of
it, all the while, so mingled with the sense of
unfathomable danger, and the human effort
and sorrow going on perpetually from age to
age, waves rolling for ever, and winds moan-
ing for ever, and faithful hearts trusting and
sickening for ever, and brave lives dashed
away about the rattling beach like weeds for
ever ; and still at the helm of every lonely
boat, through starless night and hopeless
dawn, His hand, who spread the fisher's net
over the dust of the Sidonian palaces, and
gave into the fisher's hand the keys of the
kingdom of heaven.

Next after the fishing-boat—which, as I said,
in the architecture of the sea represents the

cottage, more especially the pastoral or agri-
cultural cottage, watchful over some pathless
domain of moorland or arable, as the fishing-
boat swims humbly in the midst of the broad
green fields and hills of ocean, out of which it
has to win such fruit as they can give, and to
compass with net or drag such flocks as it
may find,—next to this ocean-cottage ranks
in interest, it seems to me, the small, over-
wrought, under-crewed, ill-caulked merchant
brig or schooner; the kind of ship which
first shows its couple of thin masts over
the low fields or marshes as we near any
third-rate seaport; and which is sure some-
where to stud the great space of glittering
water, seen from any sea-cliff, with its four
or five square-set sails. Of the larger and
more polite tribes of merchant vessels, three-
masted, and passenger-carrying, I have nothing
to say, feeling in general little sympathy with
people who want to *go* anywhere; nor caring
much about anything, which in the essence of
it expresses a desire to get to other sides of
the world; but only for homely and stay-at-
home ships, that live their life and die their
death about English rocks. Neither have I

any interest in the higher branches of commerce, such as traffic with spice islands, and
porterage of painted tea-chests or carved
ivory; for all this seems to me to fall under
the head of commerce of the drawing-room;
costly, but not venerable.   I respect in the
merchant service only those ships that carry
coals, herrings, salt, timber, iron, and such
other commodities, and that have disagreeable
odour, and unwashed decks.   But there are
few things more impressive to me than one of
these ships lying up against some lonely quay
in a black sea-fog, with the furrow traced under
its tawny keel far in the harbour slime.   The
noble misery that there is in it, the might of
its rent and strained unseemliness, its wave-
worn melancholy, resting there for a little while
in the comfortless ebb, unpitied, and claiming
no pity; still less honoured, least of all conscious
of any claim to honour; casting and craning
by due balance whatever is in its hold up to
the pier, in quiet truth of time; spinning of
wheel, and slackening of rope, and swinging
of spade, in as accurate cadence as a waltz
music; one or two of its crew, perhaps, away
forward, and a hungry boy and yelping dog

eagerly interested in something from which a
blue dull smoke rises out of pot or pan; but
dark-browed and silent, their limbs slack, like
the ropes above them, entangled as they are
in those inextricable meshes about the patched
knots and heaps of ill-reefed sable sail.   What
a majestic sense of service in all that languor !
the rest of human limbs and hearts, at utter
need, not in sweet meadows or soft air, but in
harbour slime and biting fog; so drawing their
breath once more, to go out again, without
lament, from between the two skeletons of
pier-heads, vocal with wash of under wave,
into the grey troughs of tumbling brine; there,
as they can, with slacked rope, and patched
sail, and leaky hull, again to roll and stagger
far away amidst the wind and salt sleet, from
dawn to dusk and dusk to dawn, winning day
by day their daily bread; and for last reward,
when their old hands, on some winter night,
lose feeling along the frozen ropes, and their
old eyes miss mark of the lighthouse quenched
in foam, the so-long impossible Rest, that shall
hunger no more, neither thirst any more,—their
eyes and mouths filled with the brown sea-sand.

   After these most venerable, to my mind, of

all ships, properly so styled, I find nothing of
comparable interest in any floating fabric until
we come to the great achievement of the 19th
century. For one thing this century will in
after ages be considered to have done in a
superb manner, and one thing, I think, only.
It has not distinguished itself in political
spheres ; still less in artistical. It has produced no golden age by its Reason ; neither
does it appear eminent for the constancy of
its Faith. Its telescopes and telegraphs would
be creditable to it, if it had not in their pursuit
forgotten in great part how to see clearly with
its eyes, and to talk honestly with its tongue.
Its natural history might have been creditable
to it also, if it could have conquered its habit
of considering natural history to be mainly the
art of writing Latin names on white tickets.
But, as it is, none of these things will be hereafter considered to have been got on with by
us as well as might be ; whereas it will always
be said of us, with unabated reverence,

" THEY BUILT SHIPS OF THE LINE."

Take it all in all, a Ship of the Line is the
most honourable thing that man, as a gregarious

animal, has ever produced. By himself, un-
helped, he can do better things than ships of
the line; he can make poems and pictures, and
other such concentrations of what is best in
him. But as a being living in flocks, and
hammering out, with alternate strokes and
mutual agreement, what is necessary for him
in those flocks, to get or produce, the ship of
the line is his first work. Into that he has
put as much of his human patience, common
sense, forethought, experimental philosophy,
self-control, habits of order and obedience,
thoroughly wrought handwork, defiance of
brute elements, careless courage, careful pa-
triotism, and calm expectation of the judgment of God, as can well be put into a space
of 300 feet long by 80 broad. And I am
thankful to have lived in an age when I could
see this thing so done.

Considering, then, our shipping, under the
three principal types of fishing-boat, collier,
and ship of the line, as the great glory of this
age; and the "New Forest" of mast and yard
that follows the windings of the Thames, to
be, take it all in all, a more majestic scene,
I don't say merely than any of our streets or

palaces as they now are, but even than the best that streets and palaces can generally be; it has often been a matter of serious thought to me how far this chiefly substantial thing done by the nation ought to be represented by the art of the nation; how far our great artists ought seriously to devote themselves to such perfect painting of our ships as should reveal to later generations—lost perhaps in clouds of steam and floating troughs of ashes —the aspect of an ancient ship of battle under sail.

To which, I fear, the answer must be sternly this: That no great art ever was, or can be, employed in the careful imitation of the work of man as its principal subject. That is to say, art will not bear to be reduplicated. A ship is a noble thing, and a cathedral a noble thing, but a painted ship or a painted cathedral is not a noble thing. Art which reduplicates art is necessarily second-rate art. I know no principle more irrefragably authoritative than that which I had long ago occasion to express: "All noble art is the expression of man's delight in God's work; not in his own."

"How!" it will be asked, "Are Stanfield, Isabey, and Prout necessarily artists of the second order because they paint ships and buildings instead of trees and clouds?" Yes, necessarily of the second order; so far as they paint ships rather than sea, and so far as they paint buildings rather than the natural light, and colour, and work of years upon those buildings. For, in this respect, a ruined building is a noble subject, just as far as man's work has therein been subdued by nature's; and Stanfield's chief dignity is his being a painter less of shipping than of the seal of time or decay upon shipping.* For a wrecked ship, or shattered boat, is a noble subject, while a ship in full sail, or a perfect boat, is an ignoble one; not merely because the one is by reason of its ruin more picturesque than the other, but because it is a nobler act in man to meditate upon Fate as it conquers his work, than upon that work itself.

Shipping, therefore, in its perfection, never can become the subject of noble art; and that just because to represent it in its perfection

* As in the very beautiful picture of this year's Academy, "The Abandoned."

would tax the powers of art to the utmost. If a great painter could rest in drawing a ship, as he can rest in drawing a piece of drapery, we might sometimes see vessels introduced by the noblest workmen, and treated by them with as much delight as they would show in scattering lustre over an embroidered dress, or knitting the links of a coat of mail. But ships cannot be drawn at times of rest. More complicated in their anatomy than the human frame itself, so far as that frame is outwardly discernible; liable to all kinds of strange accidental variety in position and movement, yet in each position subject to imperative laws which can only be followed by unerring knowledge; and involving, in the roundings and foldings of sail and hull, delicacies of drawing greater than exist in any other inorganic object, except perhaps a snow wreath,*—they present, irrespective of sea or sky, or anything else around them, difficulties

* The catenary and other curves of tension which a sail assumes under the united influence of the wind, its own weight, and the particular tensions of the various ropes by which it is attached, or against which it presses, show at any moment complexities of arrangement to which fidelity, except after the study of a lifetime, is impossible.

which could only be vanquished by draughts-manship quite accomplished enough to render even the subtlest lines of the human face and form. But the artist who has once attained such skill as this will not devote it to the drawing of ships. He who can paint the face of St. Paul will not elaborate the parting timbers of the vessel in which he is wrecked; and he who can represent the astonishment of the apostles at the miraculous draught will not be solicitous about accurately showing that their boat is overloaded.

"What!" it will perhaps be replied, "have, then, ships never been painted perfectly yet, even by the men who have devoted most attention to them?" Assuredly not. A ship never yet has been painted at all, in any other sense than men have been painted in "Land-scapes with figures." Things have been painted which have a general effect of ships, just as things have been painted which have a general effect of shepherds or banditti; but the best average ship-painting no more reaches the truth of ships than the equestrian troops in one of Van der Meulen's battle-pieces express the higher truths of humanity.

Take a single instance. I do not know any
work in which, on the whole, there is a more
unaffected love of ships for their own sake,
and a fresher feeling of sea breeze always
blowing, than Stanfield's "Coast Scenery."
Now, let the reader take up that book, and
look through all the plates of it at the way
in which the most important parts of a ship's
skeleton are drawn, those most wonderful
junctions of mast with mast, corresponding to
the knee or hip in the human frame, technically
known as "Tops." Under its very simplest
form, in one of those poor collier brigs, which
I have above endeavoured to recommend to
the reader's affection, the junction of the top-
gallant-mast with the topmast, when the sail
is reefed, will present itself under no less
complex and mysterious form than this in
Fig. 1, a horned knot of seven separate pieces
of timber, irrespective of the two masts and
the yard; the whole balanced and involved
in an apparently inextricable web of chain
and rope, consisting of at least sixteen ropes
about the top-gallant-mast, and some twenty-
five crossing each other in every imaginable degree of slackness and slope about the topmast.

Two-thirds of these ropes are omitted in the cut, because I could not draw them without taking more time and pains than the point

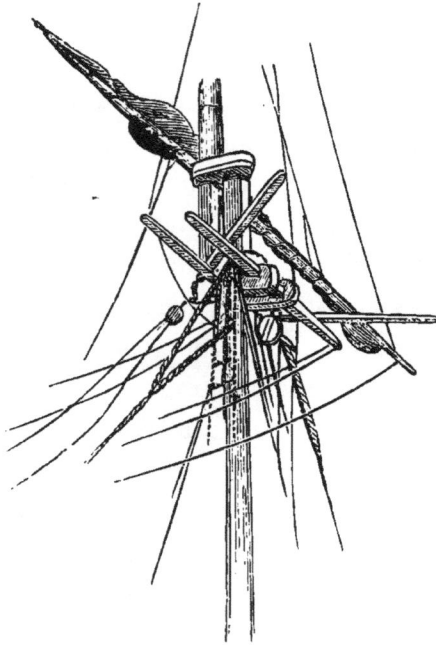

FIG. 1.

to be illustrated was worth; the thing, as it is, being drawn quite well enough to give some idea of the facts of it. Well, take up Stanfield's "Coast Scenery," and look through

it in search of tops, and you will invariably
find them represented as in Fig. 2, or even
with fewer lines; the example Fig. 2 being
one of the tops of the frigate running into
Portsmouth harbour, magnified to about twice
its size in the plate.

" Well, but it was impossible to do more on
so small a scale." By no means: but take

FIG. 2.

what scale you choose, of Stanfield's or any
other marine painter's most elaborate painting,
and let me magnify the study of the real top
in proportion, and the deficiency of detail will
always be found equally great : I mean in the
work of the higher artists, for there are of
course many efforts at greater accuracy of
delineation by those painters of ships who are

to the higher marine painter what botanical draughtsmen are to the landscapists; but just as in the botanical engraving the spirit and life of the plant are always lost, so in the technical ship-painting the life of the ship is always lost, without, as far as I can see, attaining, even by this sacrifice, anything like completeness of mechanical delineation. At least, I never saw the ship drawn yet which gave me the slightest idea of the entanglement of real rigging.

Respecting this lower kind of ship-painting, it is always matter of wonder to me that it satisfies sailors. Some years ago I happened to stand longer than pleased my pensioner guide before Turner's "Battle of Trafalgar," at Greenwich Hospital; a picture which, at a moderate estimate, is simply worth all the rest of the hospital—ground—walls—pictures and models put together. My guide, supposing me to be detained by indignant wonder at seeing it in so good a place, assented to my supposed sentiments by muttering in a low voice: "Well, sir, it *is* a shame that that thing should be there. We ought to 'a 'ad a Uggins; that's sartain." I was not surprised that my

sailor friend should be disgusted at seeing the
Victory lifted nearly right out of the water,
and all the sails of the fleet blowing about to
that extent that the crews might as well have
tried to reef as many thunder-clouds. But I
was surprised at his perfect repose of respect-
ful faith in " Uggins," who appeared to me—
unfortunate landsman as I was—to give no
more idea of the look of a ship of the line going
through the sea, than might be obtained from
seeing one of the correct models at the top of
the hall floated in a fishpond.

Leaving, however, the sailor to his enjoyment, on such grounds as it may be, of this
model drawing, and being prepared to find
only a vague and hasty shadowing forth of
shipping in the works of artists proper, we
will glance briefly at the different stages of
excellence which such shadowing forth has
reached, and note in their consecutive changes
the feelings with which shipping has been
regarded at different periods of art.

1. *Mediæval Period*. The vessel is regarded merely as a sort of sea-carriage, and
painted only so far as it is necessary for complete display of the groups of soldiers or saints

on the deck : a great deal of quaint shipping, richly hung with shields, and gorgeous with banners, is, however, thus incidently represented in 15th-century manuscripts, embedded in curly green waves of sea full of long fish ; and although there is never the slightest expression of real sea character, of motion, gloom, or spray, there is more real interest of marine detail and incident than in many later compositions.

2. *Early Venetian Period.* A great deal of tolerably careful boat-drawing occurs in the pictures of Carpaccio and Gentile Bellini, deserving separate mention among the marine schools, in confirmation of what has been stated above, that the drawing of boats is more difficult than that of the human form. For, long after all the perspectives and fore-shortenings of the human body were completely understood, as well as those of architecture, it remained utterly beyond the power of the artists of the time to draw a boat with even tolerable truth. Boats are always tilted up on end, or too long, or too short, or too high in the water. Generally they appear to be regarded with no interest whatever, and

are painted merely where they are matters of necessity. This is perfectly natural: we pronounce that there is romance in the Venetian conveyance by oars, merely because we ourselves are in the habit of being dragged by horses. A Venetian, on the other hand, sees vulgarity in a gondola, and thinks the only true romance is in a hackney coach. And thus, it was no more likely that a painter in the days of Venetian power should pay much attention to the shipping in the Grand Canal, than that an English artist should at present concentrate the brightest rays of his genius on a cab-stand.

3. *Late Venetian Period.* Deserving mention only for its notably negative character. None of the great Venetian painters, Tintoret, Titian, Veronese, Bellini, Giorgione, Bonifazio, ever introduce a ship if they can help it. They delight in ponderous architecture, in grass, flowers, blue mountains, skies, clouds, and gay dresses; nothing comes amiss to them but ships and the sea. When they are forced to introduce these, they represent merely a dark-green plain, with reddish galleys spotted about it here and there, looking much

like small models of shipping pinned on a green board. In their marine battles, there is seldom anything discernible except long rows of scarlet oars, and men in armour falling helplessly through them.

4. *Late Roman Period.* That is to say, the time of the beginning of the Renaissance landscape by the Caracci, Claude, and Salvator. First, in their landscapes, shipping begins to assume something like independent character, and to be introduced for the sake of its picturesque interest; although what interest could be taken by any healthy human creature in such vessels as were then painted has always remained a mystery to me. The ships of Claude, having hulls of a shape something between a cocoa-nut and a high-heeled shoe, balanced on their keels on the top of the water, with some scaffolding and cross-sticks above, and a flag at the top of every stick, form perhaps the *purest* exhibition of human inanity and fatuity which the arts have yet produced. The harbours also, in which these model navies ride, are worthy of all observation for the intensity of the false taste which, endeavouring to unite in them the characters

of pleasure-ground and port, destroys the veracity of both. There are many inlets of the Italian seas where sweet gardens and regular terraces descend to the water's edge; but these are not the spots where merchant vessels anchor, or where bales are disembarked. On the other hand, there are many busy quays and noisy arsenals upon the shores of Italy; but Queen's palaces are not built upon the quays, nor are the docks in any wise adorned with conservatories or ruins. It was reserved for the genius of Claude to combine the luxurious with the lucrative, and rise to a commercial ideal, in which cables are fastened to temple pillars, and lighthouses adorned with rows of beaupots. It seems strange also that any power which Salvator showed in the treatment of other subjects utterly deserts him when he approaches the sea. Though always coarse, false, and vulgar, he has at least energy, and some degree of invention, as long as he remains on land; his terrestrial atrocities are animated, and his rock-born fancies formidable. But the sea air seems to dim his sight and paralyze his hand. His love of darkness and destruction, far from

seeking sympathy in the rage of ocean, dis-
appears as he approaches the beach; after
having tortured the innocence of trees into
demoniac convulsions, and shattered the loveli-
ness of purple hills into colourless dislocation,
he approaches the real wrath and restlessness
of ocean without either admiration or dismay,
and appears to feel nothing at its shore except
a meagre interest in bathers, fishermen, and
gentlemen in court dress bargaining for state
cabins. Of all the pictures by men who bear
the reputation of great masters which I have
ever seen in my life (except only some by
Domenichino), the two large "Marines" in
the Pitti Palace, attributed to Salvator, are, on
the whole, the most vapid and vile examples
of human want of understanding. In the folly
of Claude there is still a gleam of grace and
innocence; there is refreshment in his child-
ishness, and tenderness in his inability. But
the folly of Salvator is disgusting in its very
nothingness: it is like the vacuity of a plague-
room in an hospital, shut up in uncleansed
silence, emptied of pain and motion, but not
of infection.

5. *Dutch Period.* Although in artistical

qualities lower than is easily by language ex-
pressible, the Italian marine painting usually
conveys an idea of three facts about the sea,
—that it is green, that it is deep, and that
the sun shines on it.   The dark plain which
stands for far away Adriatic with the Vene-
tians, and the glinting swells of tamed wave
which lap about the quays of Claude, agree
in giving the general impression that the
ocean consists of pure water, and is open to
the pure sky.   But the Dutch painters, while
they attain considerably greater dexterity than
the Italian in mere delineation of nautical
incident, were by nature precluded from ever
becoming aware of these common facts; and
having, in reality, never in all their lives seen
the sea, but only a shallow mixture of sea-
water and sand; and also never in all their
lives seen the sky, but only a lower element
between them and it, composed of marsh
exhalation and fog-bank; they are not to
be with too great severity reproached for the
dulness of their records of the nautical enter-
prise of Holland.   *We* only are to be reproached, who, familiar with the Atlantic, are
yet ready to accept with faith, as types of

sea, the small waves *en papillote,* and peruke-
like puffs of farinaceous foam, which were the
delight of Backhuysen and his compeers.   If
one could but arrest the connoisseurs in the
fact of looking at them with belief, and, magi-
cally introducing the image of a true sea-wave,
let it roll up to them through the room,—one
massive fathom's height and rood's breadth of
brine, passing them by but once,—dividing,
Red Sea-like, on right hand and left,—but at
least setting close before their eyes, for once
in inevitable truth, what a sea-wave really is ;
its green mountainous giddiness of wrath, its
overwhelming crest—heavy as iron, fitful as
flame, clashing against the sky in long cloven
edge,—its furrowed flanks, all ghastly clear,
deep in transparent death, but all laced across
with lurid nets of spume, and tearing open
into meshed interstices their churned veil of
silver fury, showing still the calm grey abyss
below; that has no fury and no voice, but
is as a grave always open, which the green
sighing mounds do but hide for an instant
as they pass.   Would they, shuddering back
from this wave of the true, implacable sea,
turn forthwith to the papillotes ?   It might

be so.   It is what we are all doing, more or
less, continually.

Well, let the waves go their way; it is not
of them that we have here to reason; but be
it remembered, that men who cannot enter
into the Mind of the Sea, cannot for the same
reason enter into the Mind of Ships, in their
contention with it; and the fluttering, totter-
ing, high - pooped, flag - beset fleets of these
Dutch painters have only this much superiority
over the caricatures of the Italians, that they
indeed appear in some degree to have been
studied from the high-pooped and flag-beset
nature which was in that age visible, while
the Claude and Salvator ships are ideals of
the studio.   But the effort is wholly unsuc-
cessful.   Any one who has ever attempted to
sketch a vessel in motion knows that he might
as easily attempt to sketch a bird on the wing,
or a trout on the dart.   Ships can only be
drawn, as animals must be, by the high instinct of momentary perception, which rarely
developed itself in any Dutch painter, and
least of all in their painters of marine.   And
thus the awkward forms of shipping, the
shallow impurity of the sea, and the cold

incapacity of the painter, joining in disadvantageous influence over them, the Dutch marine paintings may be simply, but circumstantially, described as the misrepresentation of undeveloped shipping in a discoloured sea by distempered painters. An exception ought to be made in favour of the boats of Cuyp, which are generally well floated in calm and sunny water; and, though rather punts or tubs than boats, have in them some elements of a slow, warm, square-sailed, sleepy grandeur—respectable always, when compared either with the flickering follies of Backhuysen, or the monstrous, unmanly, and, *à fortiori*, unsailorly absurdities of metaphysical vessels, puffed on their way by corpulent genii, or pushed by protuberant dolphins, which Rubens and the other so-called historical painters of his time were accustomed to introduce in the mythology of their court-adulation; that marvellous Faith of the 18th century, which will one day, and that not far off, be known for a thing more truly disgraceful to human nature than the Polynesian's dance round his feather idol, or Egyptian's worship of the food he fattened on. From Salvator and Domenichino it is possible to turn in a proud indignation, knowing that theirs are no fair examples of the human mind; but it is with humbled and woful anger that we must trace the degradation of the intellect of Rubens in his pictures of the life of Mary of Medicis.*

6. *Modern Period.* The gradual appreciation of the true character both of shipping and the ocean, in the works of the painters of

---

* " The town of Lyons, seated upon a chariot drawn by two lions, *lifts its eyes towards heaven*, and admires there— ' les nouveaux Epoux,'—represented in the character of Jupiter and Juno."—*Notice des Tableaux du Musée Impérial*, 2nde partie, Paris, 1854, p. 235.

"The Queen upon her throne holds with one hand the sceptre, in the other the balance. Minerva and Cupid are at her sides. Abundance and Prosperity distribute medals, laurels, 'et d'autres récompenses,' to the Genii of the Fine Arts. Time, crowned with the productions of the seasons, leads France to the—Age of Gold ! "—p. 239.

So thought the Queen, and Rubens, and the Court. Time himself, "crowned with the productions of the seasons," was, meanwhile, as Thomas Carlyle would have told us, "quite of another opinion."

With view of arrival at Golden Age all the sooner, the Court determine to go by water ; "and Marie de Medicis gives to her son the government of the state, under the emblem of a vessel, of which he holds the rudder."

This piece of royal pilotage, being on the whole the most characteristic example I remember of the Mythological marine above alluded to, is accordingly recommended to the reader's serious attention.

the last half century, is part of that successful
study of other elements of landscape, of which
I have long laboured at a consistent investiga-
tion, now partly laid before the public; I shall
not, therefore, here enter into any general
inquiry respecting modern sea-painting, but
limit myself to a notice of the particular feelings which influenced Turner in his marine
studies, so far as they are shown in the series
of plates which have now been trusted to me
for illustration.

Among the earliest sketches from nature
which Turner appears to have made, in pencil
and Indian ink, when a boy of twelve or
fourteen, it is very singular how large a proportion consists of careful studies of stranded
boats. Now, after some fifteen years of
conscientious labour, with the single view of
acquiring knowledge of the ends and powers
of art, I have come to one conclusion, which at
the beginning of those fifteen years would have
been very astonishing to myself—that, of all
our modern school of landscape painters, next
to Turner, and before the rise of the Pre-
Raphaelites, the man whose works are on the
whole most valuable, and show the highest

intellect, is Samuel Prout. It is very notable
that also in Prout's early studies, shipping
subjects took not merely a prominent, but I
think even a principal, place.

The reason of this is very evident: both
Turner and Prout had in them an untaught,
inherent perception of what was great and
pictorial. They could not find it in the buildings or in the scenes immediately around
them. But they saw some element of real
power in the boats. Prout afterwards found
material suited to his genius in other direc-
tions, and left his first love; but Turner
retained the early affection to the close of
his life, and the last oil picture which he
painted, before his noble hand forgot its
cunning, was the Wreck - buoy. The last
thoroughly perfect picture he ever painted,
was the Old Téméraire.

The studies which he was able to make
from nature in his early years, are chiefly of
fishing-boats, barges, and other minor marine
still life; and his better acquaintance with this
kind of shipping than with the larger kind is
very marked in the Liber Studiorum, in which
there are five careful studies of fishing-boats

under various circumstances; namely, Calais Harbour, Sir John Mildmay's Picture, Flint Castle, Marine Dabblers, and the Calm; while of other shipping, there are only two subjects, both exceedingly unsatisfactory.

Turner, however, deemed it necessary to his reputation at that period that he should paint pictures in the style of Vandevelde; and, in order to render the resemblance more complete, he appears to have made careful drawings of the different parts of old Dutch shipping. I found a large number of such drawings among the contents of his neglected portfolios at his death; some were clearly not by his own hand, others appeared to be transcripts by him from prints or earlier drawings; the quantity altogether was very great, and the evidence of his prolonged attention to the subject more distinct than with respect to any other element of landscape. Of plants, rocks, or architecture, there were very few careful pieces of anatomical study. But several drawers were entirely filled with these memoranda of shipping.

In executing the series of drawings for the work known as the Southern Coast, Turner appears to have gained many ideas about shipping, which, once received, he laid up by him for use in after years. The evidence of this laying by of thought in his mind, as it were in reserve, until he had power to express it, is curious and complete throughout his life; and although the Southern Coast drawings are for the most part quiet in feeling, and remarkably simple in their mode of execution, I believe it was in the watch over the Cornish and Dorsetshire coast, which the making of those drawings involved, that he received all his noblest ideas about sea and ships.

Of one thing I am certain; Turner never drew anything that could be *seen*, without having seen it. That is to say, though he would draw Jerusalem from some one else's sketch, it would be, nevertheless, entirely from his own experience of ruined walls: and though he would draw ancient shipping (for an imitation of Vandevelde, or a vignette to the voyage of Columbus) from such data as he could get about things which he could no more see with his own eyes, yet when, of his own free will, in the subject of Ilfracombe, he, in the year 1818, introduces a shipwreck, I am perfectly certain that, before the year 1818, he had *seen* a shipwreck, and, moreover, one of that horrible kind—a ship dashed to pieces in deep water, at the foot of an inaccessible cliff. Having once seen this, I perceive, also, that the image of it could not be effaced from his mind. It taught him two great facts, which he never afterwards forgot; namely, that both ships and sea were things that broke to pieces. *He never afterwards painted a ship quite in fair order.* There is invariably a feeling about his vessels of strange awe and danger; the sails are in some way loosening, or flapping as if in fear; the swing of the hull, majestic as it may be, seems more at the mercy of the sea than in triumph over it; the ship never looks gay, never proud, only warlike and enduring. The motto he chose, in the Catalogue of the Academy, for the most cheerful marine he ever painted, the Sun of Venice going to Sea, marked the uppermost feeling in his mind:

> " Nor heeds the Demon that in grim repose
> Expects his evening prey."

I notice above the subject of his last marine

D

picture, the Wreck-buoy, and I am well per-
suaded that from that year 1818, when first
he saw a ship rent asunder, he never beheld
one at sea, without, in his mind's eye, at the
same instant, seeing her skeleton.

But he had seen more than the death of the
ship.   He had seen the sea feed her white
flames on souls of men; and heard what a
storm-gust sounded like, that had taken up
with it, in its swirl of a moment, the last
breaths of a ship's crew.   He never forgot
either the sight or the sound.   Among the
last plates prepared by his own hand for the
Liber Studiorum, (all of them, as was likely
from his advanced knowledge, finer than any
previous pieces of the series, and most of
them unfortunately never published, being
retained beside him for some last touch—for
ever delayed,) perhaps the most important is
one of the body of a drowned sailor, dashed
against a vertical rock in the jaws of one
merciless, immeasurable wave.   He repeated
the same idea, though more feebly expressed,
later in life, in a small drawing of Grandville,
on the coast of France.   The sailor clinging
to the boat in the marvellous drawing of

Dunbar is another reminiscence of the same kind. He hardly ever painted a steep rocky coast without some fragment of a devoured ship, grinding in the blanched teeth of the surges,—just enough left to be a token of utter destruction. Of his two most important paintings of definite shipwreck I shall speak presently.

I said that at this period he first was assured of another fact, namely, that the *Sea* also was a thing that broke to pieces. The sea up to that time had been generally regarded by painters as a liquidly composed, level-seeking consistent thing, with a smooth surface, rising to a water-mark on sides of ships; in which ships were scientifically to be embedded, and wetted, up to said water-mark, and to remain dry above the same. But Turner found during his Southern Coast tour that the sea was *not* this: that it was, on the contrary, a very incalculable and unhorizontal thing, setting its "water mark" sometimes on the highest heavens, as well as on sides of ships;—very breakable into pieces; half of a wave separable from the other half, and on the instant carriageable miles inland;—not in any wise limiting itself to a state of apparent liquidity, but now striking like a steel gauntlet, and now becoming a cloud, and vanishing, no eye could tell whither; one moment a flint cave, the next a marble pillar, the next a mere white fleece thickening the thundery rain. He never forgot those facts; never afterwards was able to recover the idea of positive distinction between sea and sky, or sea and land. Steel gauntlet, black rock, white cloud, and men and masts gnashed to pieces and disappearing in a few breaths and splinters among them;—a little blood on the rock angle, like red sea-weed, spunged away by the next splash of the foam, and the glistering granite and green water all pure again in vacant wrath. So stayed by him, for ever, the Image of the Sea.

One effect of this revelation of the nature of ocean to him was not a little singular. It seemed that ever afterwards his appreciation of the calmness of water was deepened by what he had witnessed of its frenzy, and a certain class of entirely tame ·subjects were treated by him even with increased affection after he had seen the full manifestation of sublimity. He had always a great regard for

canal boats, and instead of sacrificing these
old, and one would have thought unentertain-
ing, friends to the deities of Storm, he seems
to have returned with a lulling pleasure from
the foam and danger of the beach to the sedgy
bank and stealthy barge of the lowland river.
Thenceforward his work which introduces
shipping is divided into two classes; one
embodying the poetry of silence and calmness,
the other of turbulence and wrath. Of inter-
mediate conditions he gives few examples; if
he lets the wind down upon the sea at all,
it is nearly always violent, and though the
waves may not be running high, the foam is
torn off them in a way which shows they will
soon run higher. On the other hand, nothing
is so perfectly calm as Turner's calmness. To
the canal barges of England he soon added
other types of languid motion ; the broad-
ruddered barques of the Loire, the droop-
ing sails of Seine, the arcaded barques of
the Italian lakes slumbering on expanse of
mountain-guarded wave, the dreamy prows of
pausing gondolas on lagoons at moon-rise ; in
each and all commanding an intensity of calm,
chiefly because he never admitted an instant's

rigidity.   The surface of quiet water with
other painters becomes FIXED.   With Turner
it looks as if a fairy's breath would stir it,
but the fairy's breath is not there.   So also
his boats are intensely motionless, because intensely capable of motion.   No other painter
ever floated a boat quite rightly; all other
boats stand on the water, or are fastened in
it; only his *float* in it.   It is very difficult to
trace the reasons of this, for the rightness of
the placing on the water depends on such
subtle curves and shadows in the floating
object and its reflection, that in most cases the
question of entirely right or entirely wrong
resolves itself into the "estimation of an
hair": and what makes the matter more difficult still, is, that sometimes we may see a
boat drawn with the most studied correctness
in every part, which yet will not swim; and
sometimes we may find one drawn with many
easily ascertainable errors, which yet swims
well enough; so that the drawing of boats is
something like the building of them, one may
set off their lines by the most authentic rules,
and yet never be sure they will sail well.
It is, however, to be observed that Turner

seemed, in those southern coast storms, to
have been somewhat too strongly impressed
by the disappearance of smaller crafts in surf,
and was wont afterwards to give an uncom-
fortable aspect even to his gentlest seas, by
burying his boats too deeply. When he
erred, in this or other matters, it was not
from want of pains, for of all accessories to
landscape, ships were throughout his life those
which he studied with the greatest care. His
figures, whatever their merit or demerit, are
certainly never the beloved part of his work ;
and though the architecture was in his early
drawings careful, and continued to be so down
to the Hakewell's Italy series, it soon became
mannered and false whenever it was principal.
He would indeed draw a ruined tower, or a
distant town, incomparably better than any
one else, and a staircase or a bit of balustrade
very carefully ; but his temples and cathe-
drals showed great ignorance of detail, and
want of understanding of their character. But
I am aware of no painting from the beginning
of his life to its close, containing *modern*
shipping as its principal subject, in which he
did not put forth his full strength, and pour

out his knowledge of detail with a joy which renders those works, as a series, among the most valuable he ever produced.   Take for instance :

    1. Lord Yarborough's Shipwreck.
    2. The Trafalgar, at Greenwich Hospital.
    3. The Trafalgar, in his own gallery.
    4. The Pas de Calais.
    5. The Large Cologne.
    6. The Havre.
    7. The Old Téméraire.

I know no fourteen pictures by Turner for which these seven might be wisely changed; and in all of these the shipping is thoroughly principal, and studied from existing ships.   A large number of inferior works were, however, also produced by him in imitation of Vandevelde, representing old Dutch shipping; in these the shipping is scattered, scudding and distant, the sea grey and lightly broken. Such pictures are, generally speaking, among those of least value which he has produced. Two very important ones, however, belong to the imitative school : Lord Ellesmere's, founded on Vandevelde; and the Dort, at Farnley, on

Cuyp.   The latter, as founded on the better
master, is the better picture, but still pos-
sesses few of the true Turner qualities, except
his peculiar calmness, in which respect it is un-
rivalled ; and if joined with Lord Yarborough's
Shipwreck, the two may be considered as the
principal symbols, in Turner's early oil paintings, of his two strengths in Terror and
Repose.   Among his drawings, shipping, as
the principal subject, does not always constitute a work of the first class ; nor does it so
often occur.   For the difficulty, in a drawing,
of getting good colour is so much less, and
that of getting good form so much greater,
than in oil, that Turner naturally threw his
elaborate studies of ship form into oil, and
made his noblest work in drawing rich in
hues of landscape.   Yet the Cowes, Devonport, and Gosport, from the England and
Wales (the Saltash is an inferior work), united
with two drawings of this series, Portsmouth
and Sheerness, and two from Farnley, one of
the wreck of an Indiaman, and the other of a
ship of the line taking stores, would form a
series, not indeed as attractive at first sight
as many others, but embracing perhaps more

of Turner's peculiar, unexampled, and unapproachable gifts than any other group of drawings which could be selected, the choice being confined to one class of subject.

I have only to state, in conclusion, that these twelve drawings of the Harbours of England are more representable by engraving than most of his works. Few parts of them are brilliant in colour; they were executed chiefly in brown and blue, and with more direct reference to the future engraving than was common with Turner. They are also small in size, generally of the exact dimensions of the plate, and therefore the lines of the compositions are not spoiled by contraction; while finally, the touch of the painter's hand upon the wave-surface is far better imitated by mezzotint engraving than by any of the ordinary expedients of line. Take them all in all, they form the most valuable series of marine studies which have as yet been published from his works; and I hope that they may be of some use hereafter in recalling the ordinary aspect of our English seas, at the exact period when the nation had done its utmost in the wooden and woven strength of ships, and had most perfectly fulfilled the old and noble prophecy—

> " They shall ride
> Over ocean wide,
> With hempen bridle, and horse of tree."
>
> *Thomas of Ercildoune.*

DOVER

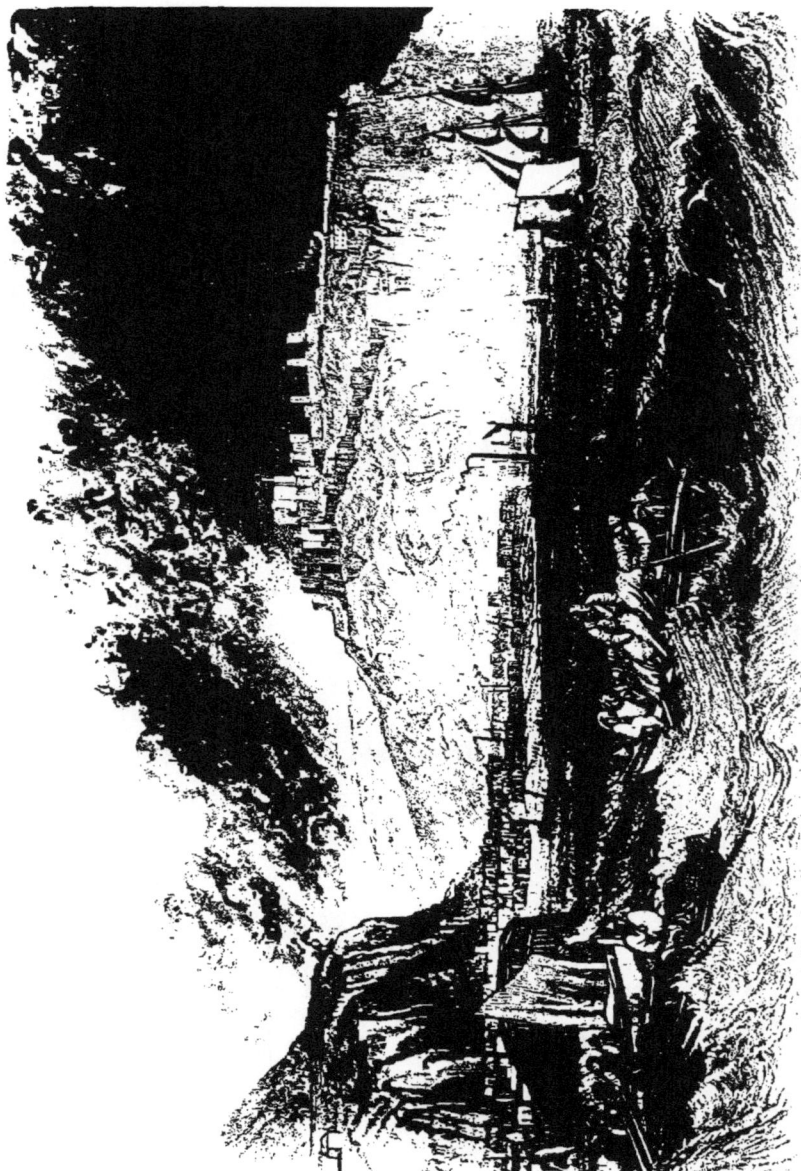

I.—DOVER

THIS port has some right to take precedence of others, as being that assuredly which first exercises the hospitality of England to the majority of strangers who set foot on her shores. I place it first therefore among our present subjects; though the drawing itself, and chiefly on account of its manifestation of Turner's faulty habit of local exaggeration, deserves no such pre-eminence. He always painted, not the place itself, but his impression of it, and this on steady principle; leaving to inferior artists the task of topographical detail; and he was right in this principle, as I have shown elsewhere, when the impression was a genuine one; but in the present case it is not so. He has lost the real character of Dover Cliffs by making the town at their feet three times lower in proportionate height than it really is; nor is he to be justified in giving the barracks, which appear on the left hand, more the air of a hospice on the top of an Alpine precipice, than of an establishment which, out of Snargate street, can be reached, without drawing breath, by a winding stair of some 170 steps ; making the slope beside them more like the side of Skiddaw than what it really is, the earthwork of an unimportant battery.

This design is also remarkable as an instance of that restlessness which was above noticed even in Turner's least stormy seas. There is nothing tremendous here in scale of wave, but the whole surface is fretted and disquieted by torturing wind ; an effect which was always increased during the progress of the subjects, by Turner's habit of scratching out small sparkling lights, in order to make the plate " bright," or "lively." * In a general way the engravers used to like this, and, as far as they were able, would tempt Turner farther into the practice, which was precisely equivalent to that of supplying the place of healthy and heart-whole cheerfulness by dram-drinking.

* See the farther explanation of this practice in the notice of the subject of " Portsmouth."

The two seagulls in the front of the picture were additions of this kind, and are
very injurious, confusing the organisation and
concealing the power of the sea. The merits
of the drawing are, however, still great as a
piece of composition. The left-hand side is
most interesting, and characteristic of Turner :
no other artist would have put the round pier
so exactly under the round cliff. It is under
it so accurately, that if the nearly vertical
falling line of that cliff be continued, it strikes
the sea-base of the pier to a hair's breadth.
But Turner knew better than any man the
value of echo, as well as of contrast,—of
repetition, as well as of opposition. The
round pier repeats the line of the main cliff,
and then the sail repeats the diagonal shadow
which crosses it, and emerges above it just
as the embankment does above the cliff brow.
Lower, come the opposing curves in the
two boats, the whole forming one group of
sequent lines up the whole side of the picture.
The rest of the composition is more common-
place than is usual with the great master ;
but there are beautiful transitions of light
and shade between the sails of the little

E

fishing-boat, the brig behind her, and the cliffs. Note how dexterously the two front sails * of the brig are brought on the top of the white sail of the fishing-boat to help to detach it from the white cliffs.

* I think I shall be generally more intelligible by ex-plaining what I mean in this way, and run less chance of making myself ridiculous in the eyes of sensible people, than by displaying the very small nautical knowledge I possess. My sailor friends will perhaps be gracious enough to believe that I *could* call these sails by their right names if I liked.

RAMSGATE

II.—RAMSGATE

THIS, though less attractive, at first sight, than the former plate, is a better example of the master, and far truer and nobler as a piece of thought. The lifting of the brig on the wave is very daring; just one of the things which is seen in every gale, but which no other painter than Turner ever represented; and the lurid transparency of the dark sky, and wild expression of wind in the fluttering of the falling sails of the vessel running into the harbour, are as fine as anything of the kind he has done. There is great grace in the drawing of this latter vessel : note the delicate switch forward of her upper mast.

There is a very singular point connected with the composition of this drawing, proving it (as from internal evidence was most likely) to be a record of a thing actually seen.

Three years before the date of this engraving Turner had made a drawing of Ramsgate for the Southern Coast series. That drawing represents the *same day*, the *same moment*, and the *same ships*, from a different point of view. It supposes the spectator placed in a boat some distance out at sea, beyond the fishing-boats on the left in the present plate, and looking towards the town, or into the harbour. The brig, which is near us here, is then, of course, in the distance on the right; the schooner entering the harbour, and, in both plates, lowering her fore-topsail, is, of course, seen foreshortened; the fishing-boats only are a little different in position and set of sail. The sky is precisely the same, only a dark piece of it, which is too far to the right to be included in *this* view, enters into the wider distance of the other, and the town, of course, becomes a more important object.

The persistence in one conception furnishes evidence of the very highest imaginative power. On a common mind, what it has seen is so feebly impressed, that it mixes other ideas with it immediately; forgets it

—modifies it—adorns it,—does anything but keep *hold* of it. But when Turner had once seen that stormy hour at Ramsgate harbour-mouth, he never quitted his grasp of it. He had *seen* the two vessels; one go in, the other out. He could have only seen them at that one moment—from one point; but the impression on his imagination is so strong, that he is able to handle it three years after-wards, as if it were a real thing, and turn it round on the table of his brain, and look at it from the other corner. He will see the brig near, instead of far off: set the whole sea and sky so many points round to the south, and see how they look, so. I never traced power of this kind in any other man.

PLYMOUTH

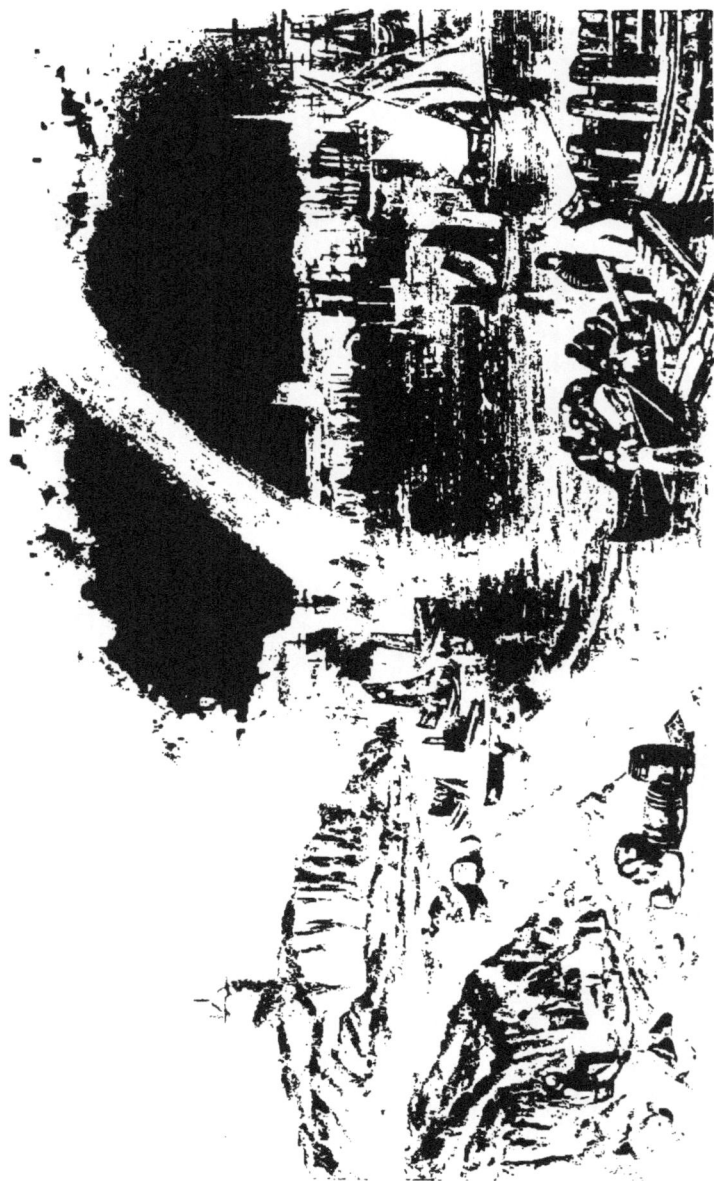

III.—PLYMOUTH

THE drawing for this plate is one of Turner's most remarkable, though not most meritorious, works : it contains the brightest rainbow he ever painted, to my knowledge ; not the best, but the most dazzling. It has been much modified in the plate. It is very like one of Turner's pieces of caprice to introduce a rainbow at all as a principal feature in such a scene ; for it is not through the colours of the iris that we generally expect to be shown eighteen-pounder batteries and ninety-gun ships.

Whether he meant the dark cloud (intensely dark blue in the original drawing), with the sunshine pursuing it back into distance ; and the rainbow, with its base set on a ship of battle, to be together types of war and peace, and of the one as the foundation of the other, I leave it to the reader to decide. My own impression is, that although Turner might have some askance symbolism in his mind, the present design is, like the former one, in many points a simple reminiscence of a seen fact.*

However, whether reminiscent or symbolic, the design is, to my mind, an exceedingly unsatisfactory one, owing to its total want of principal subject. The fort ceases to be of importance because of the bank and tower in front of it; the ships, necessarily for the effect, but fatally for themselves, are confused, and incompletely drawn, except the little sloop, which looks paltry and like a toy; and the foreground objects are, for work of Turner, curiously ungraceful and uninteresting.

It is possible, however, that to some minds

---

* I have discovered, since this was written, that the design was made from a vigorous and interesting sketch by Mr. S. Cousins, in which the rainbow and most of the ships are already in their places. Turner was, therefore, in this case, as I have found him in several other instances, realising, not a fact seen by himself, but a fact as he supposed it to have been seen by another.

the fresh and dewy space of darkness, so
animated with latent human power, may give
a sensation of great pleasure, and at all
events the design is worth study on account
of its very strangeness.

CATWATER

IV.—CATWATER

I HAVE placed in the middle of the series those pictures which I think least interesting, though the want of interest is owing more to the monotony of their character than to any real deficiency in their subjects. If, after contemplating paintings of arid deserts or glowing sunsets, we had come suddenly upon this breezy entrance to the crowded cove of Plymouth, it would have gladdened our hearts to purpose; but having already been at sea for some time, there is little in this drawing to produce renewal of pleasurable impression: only one useful thought may be gathered from the very feeling of monotony. At the time when Turner executed these drawings, his portfolios were full of the most magnificent subjects—coast and inland,—gathered from all the noblest scenery of France and Italy. He was ready

to realise these sketches for any one who
would have asked it of him, but no consistent
effort was ever made to call forth his powers;
and the only means by which it was thought
that the public patronage could be secured
for a work of this kind, was by keeping
familiar names before the eye, and awakening
the so-called " patriotic," but in reality narrow
and selfish, associations belonging to well-
known towns or watering-places.   It is to be
hoped, that when a great landscape painter
appears among us again, we may know better
how to employ him, and set him to paint for
us things which are less easily seen, and
which are somewhat better worth seeing,
than the mists of the Catwater, or terraces
of Margate.

SHEERNESS

V.—SHEERNESS

I LOOK upon this as one of the noblest sea-pieces which Turner ever produced. It has not his usual fault of over-crowding or over-glitter; the objects in it are few and noble, and the space infinite. The sky is quite one of his best: not violently black, but full of gloom and power; the complicated roundings of its volumes behind the sloop's mast, and downwards to the left, have been rendered by the engraver with notable success; and the dim light entering along the horizon, full of rain, behind the ship of war, is true and grand in the highest degree. By comparing it with the extreme darkness of the skies in the Plymouth, Dover, and Ramsgate, the reader will see how much more majesty there is in moderation than in extravagance, and how much more darkness, as far as sky is concerned, there is in grey than in black. It

85

is not that the Plymouth and Dover skies are
false,—such impenetrable forms of thunder-
cloud are amongst the commonest phenomena
of storm; but they have more of spent flash
and past shower in them than the less passionate, but more truly stormy and threaten-
ing, volumes of the sky here. The Plymouth
storm will very thoroughly wet the sails, and
wash the decks, of the ships at anchor, but
will send nothing to the bottom. For these
pale and lurid masses, there is no saying
what evil they may have in their thoughts,
or what they may have to answer for before
night. The ship of war in the distance is
one of many instances of Turner's dislike to
draw *complete* rigging; and this not only
because he chose to give an idea of his
ships having seen rough service, and being
crippled; but also because in men-of-war he
liked the mass of the hull to be increased in
apparent weight and size by want of upper
spars. All artists of any rank share this
last feeling. Stanfield never makes a careful
study of a hull without shaking some or all
of its masts out of it first, if possible. See,
in the Coast Scenery, Portsmouth harbour,

Falmouth, Hamoaze, and Rye old harbours;
and compare, among Turner's works, the
near hulls in the Devonport, Saltash, and
Castle Upnor, and distance of Gosport. The
fact is, partly that the precision of line in the
complete spars of a man-of-war is too formal
to come well into pictorial arrangements, and
partly that the chief glory of a ship of the
line is in its aspect of being "one that hath
had losses."

The subtle varieties of curve in the drawing of the sails of the near sloop are altogether exquisite; as well as the contrast of
her black and glistering side with those sails,
and with the sea. Examine the wayward
and delicate play of the dancing waves along
her flank, and between her and the brig in
ballast, plunging slowly before the wind; I
have not often seen anything so perfect in
fancy, or in execution of engraving.

The heaving and black buoy in the near
sea is one of Turner's "echoes," repeating,
with slight change, the head of the sloop
with its flash of lustre. The chief aim of
this buoy is, however, to give comparative
lightness to the shadowed part of the sea,

which is, indeed, somewhat overcharged in darkness, and would have been felt to be so, but for this contrasting mass. Hide it with the hand, and this will be immediately felt. There is only one other of Turner's works which, in its way, can be matched with this drawing, namely, the Mouth of the Humber in the River Scenery. The latter is, on the whole, the finer picture; but this by much the more interesting in the shipping.

MARGATE

VI.—MARGATE

This plate is not, at first sight, one of the most striking of the series; but it is very beautiful, and highly characteristic of Turner.*
First, in its choice of subjects: for it seems very notably capricious in a painter eminently capable of rendering scenes of sublimity and mystery, to devote himself to the delineation of one of the most prosaic of English watering-places — not once or twice, but in a series of elaborate drawings, of which this is the fourth. The first appeared in the Southern Coast series, and was followed by an elaborate drawing on a large scale, with a beautiful sunrise; then came another careful and very beautiful drawing in the England and Wales series; and finally this, which is

* It was left unfinished at his death, and I would not allow it to be touched afterwards, desiring that the series should remain as far as possible in an authentic state.

a sort of poetical abstract of the first.   Now,
if we enumerate the English ports one by
one, from Berwick to Whitehaven, round the
island, there will hardly be found another so
utterly devoid of all picturesque or romantic
interest as Margate.   Nearly all have some
steep eminence of down or cliff, some pretty
retiring dingle, some roughness of old harbour
or straggling fisher-hamlet, some fragment
of castle or abbey on the heights above,
capable of becoming a leading point in a
picture ; but Margate is simply a mass of
modern parades and streets, with a little bit
of chalk cliff, an orderly pier, and some
bathing-machines.   Turner never conceives
it as anything else ; and yet for the sake of
this simple vision, again and again he quits
all higher thoughts.   The beautiful bays of
Northern Devon and Cornwall he never
painted but once, and that very imperfectly.
The finest subjects of the Southern Coast
series—the Minehead, Clovelly, Ilfracombe,
Watchet, East and West Looe, Tintagel,
Boscastle—he never touched again ; but he
repeated Ramsgate, Deal, Dover, and Margate,
I know not how often.

Whether his desire for popularity, which, in spite of his occasional rough defiances of public opinion, was always great, led him to the selection of those subjects which he thought might meet with most acceptance from a large class of the London public, or whether he had himself more pleasurable associations connected with these places than with others, I know not; but the fact of the choice itself is a very mournful one, considered with respect to the future interests of art. There is only this one point to be remembered, as tending to lessen our regret, that it is possible Turner might have felt the necessity of compelling himself sometimes to dwell on the most familiar and prosaic scenery, in order to prevent his becoming so much accustomed to that of a higher class as to diminish his enthusiasm in its presence. Into this probability I shall have occasion to examine at greater length hereafter.

The plate of Margate now before us is nearly as complete a duplicate of the Southern Coast view as the previous plate is of that of Ramsgate; with this difference, that the position of the spectator is here the same, but the class of ship is altered, though the ship remains precisely in the same spot. A piece of old wreck, which was rather an important object to the left of the other drawing, is here removed. The figures are employed in the same manner in both designs.

The details of the houses of the town are executed in the original drawing with a precision which adds almost painfully to their natural formality. It is certainly provoking to find the great painter, who often only deigns to bestow on some Rhenish fortress or French city, crested with Gothic towers, a few misty and indistinguishable touches of his brush, setting himself to indicate, with unerring toil, every separate square window in the parades, hotels, and circulating libraries of an English bathing-place.

The whole of the drawing is well executed, and free from fault or affectation, except perhaps in the somewhat confused curlings of the near sea. I had much rather have seen it breaking in the usual straightforward way. The brilliant white of the piece of chalk cliff is evidently one of the principal aims of the composition. In the drawing the sea is throughout of a dark fresh blue, the sky greyish blue, and the grass on the top of the cliffs a little sunburnt, the cliffs themselves being left in the almost untouched white of the paper.

PORTSMOUTH

G

VII.—PORTSMOUTH

THIS beautiful drawing is a *third* recurrence
by Turner to his earliest impression of Portsmouth, given in the Southern Coast series.
The buildings introduced differ only by a
slight turn of the spectator towards the right;
the buoy is in the same spot; the man-of-
war's boat nearly so; the sloop exactly so,
but on a different tack; and the man-of-war,
which is far off to the left at anchor in the
Southern Coast view, is here nearer, and
getting up her anchor.

The idea had previously passed through
one phase of greater change, in his drawing
of "Gosport" for the England, in which,
while the sky of the Southern Coast view
was almost cloud for cloud retained, the
interest of the distant ships of the line
had been divided with a collier brig and a
fast-sailing boat. In the present view he

returns to his early thought, dwelling, however, now with chief insistence on the ship
of the line, which is certainly the most
majestic of all that he has introduced in his
drawings.

It is also a very curious instance of that
habit of Turner's before referred to (p. 49),
of never painting a ship quite in good order.
On showing this plate the other day to a
naval officer, he complained of it, first that
"the jib * would not be wanted with the
wind blowing out of harbour," and, secondly,
that " a man-of-war would never have her
foretop-gallant sail set, and her main and
mizen top-gallants furled :—all the men
would be on the yards at once."

I believe this criticism to be perfectly just,
though it has happened to me, very singularly, whenever I have had the opportunity
of making complete inquiry into any technical
matter of this kind, respecting which some
professional person had blamed Turner, that
I have always found, in the end, Turner
was right, and the professional critic wrong,

* The sail seen, edge on, like a white sword, at the head
of the ship.

owing to some want of allowance for possible
accidents, and for necessary modes of pictorial representation.   Still, this cannot be
the case in every instance ; and supposing
my sailor informant to be perfectly right in
the present one, the disorderliness of the
way in which this ship is represented as
setting her sails, gives us farther proof of
the imperative instinct in the artist's mind,
refusing to contemplate a ship, even in her
proudest moments, but as in some way over-
mastered by the strengths of chance and
storm.

The wave on the left hand beneath the
buoy, presents a most interesting example of
the way in which Turner used to spoil his
work by retouching.   All his truly fine drawings are either done quickly, or at all events
straight forward, without alteration : he never,
as far as I have examined his works hitherto,
altered but to destroy.   When he saw a
plate look somewhat dead or heavy, as, compared with the drawing, it was almost sure
at first to do, he used to scratch out little
lights all over it, and make it " sparkling " ;
a process in which the engravers almost

unanimously delighted,* and over the im-
possibility of which they now mourn, declar-
ing it to be hopeless to engrave after Turner,
since he cannot now scratch their plates for
them.   It is quite true that these small lights
were always placed beautifully; and though
the plate, after its "touching," generally
looked as if ingeniously salted out of her
dredging-box by an artistical cook, the salt-
ing was done with a spirit which no one else
can now imitate.   But the original power of
the work was for ever destroyed.   If the
reader will look carefully beneath the white
touches on the left in this sea, he will dis-
cern dimly the form of a round nodding hollow
breaker.   This in the early state of the plate
is a gaunt, dark, angry wave, rising at the
shoal indicated by the buoy;—Mr. Lupton
has fac-similed with so singular skill the
scratches of the penknife by which Turner
afterwards disguised this breaker, and spoiled
his picture, that the plate in its present state
is almost as interesting as the touched proof

---

* Not, let me say with all due honour to him, the careful
and skilful engraver of these plates, who has been much
more tormented than helped by Turner's alterations.

itself; interesting, however, only as a warn-
ing to all artists never to lose hold of their
first conception. They may tire even of
what is exquisitely right, as they work it out,
and their only safety is in the self-denial of
calm completion.

FALMOUTH

VIII.—FALMOUTH

THIS is one of the most beautiful and best-finished plates of the series, and Turner has taken great pains with the drawing; but it is sadly open to the same charges which were brought against the Dover, of an attempt to reach a false sublimity by magnifying things in themselves insignificant. The fact is that Turner, when he prepared these drawings, had been newly inspired by the scenery of the Continent; and with his mind entirely occupied by the ruined towers of the Rhine, he found himself called upon to return to the formal embrasures and unappalling elevations of English forts and hills. But it was impossible for him to recover the simplicity and narrowness of conception in which he had executed the drawing of the Southern Coast, or to regain the innocence of delight with

which he had once assisted gravely at the
drying of clothes over the limekiln at Comb
Martin, or pencilled the woodland outlines of
the banks of Dartmouth Cove. In certain
fits of prosaic humourism, he would, as we
have seen, condemn himself to delineation
of the parades of a watering-place; but the
moment he permitted himself to be enthusiastic, vaster imaginations crowded in upon
him : to modify his old conception in the
least, was to exaggerate it ; the mount of
Pendennis is lifted into rivalship with Ehren-
breitstein, and hardworked Falmouth glitters
along the distant bay, like the gay magnifi-
cence of Resina or Sorrento.

This effort at sublimity is all the more to
be regretted, because it never succeeds completely.   Shade, or magnify, or mystify as
he may, even Turner cannot make the minute
neatness of the English fort appeal to us as
forcibly as the remnants of Gothic wall and
tower that crown the Continental crags ; and
invest them as he may with smoke or sun-
beam, the details of our little mounded hills
will not take the rank of cliffs of Alp, or
promontories of Apennine ; and we lose the

English simplicity, without gaining the Continental nobleness.

I have also a prejudice against this picture for being disagreeably noisy. Wherever there is something serious to be done, as in a battle piece, the noise becomes an element of the sublimity; but to have great guns going off in every direction beneath one's feet on the right, and all round the other side of the castle, and from the deck of the ship of the line, and from the battery far down the cove, and from the fort on the top of the hill, and all for nothing, is to my mind eminently troublesome.

The drawing of the different wreaths and depths of smoke, and the explosive look of the flash on the right, are, however, very wonderful and peculiarly Turneresque ; the sky is also beautiful in form, and the foreground, in which we find his old regard for washerwomen has not quite deserted him, singularly skilful. It is curious how formal the whole picture becomes if this figure and the grey stones beside it are hidden with the hand.

SIDMOUTH

IX.—SIDMOUTH

THIS drawing has always been interesting to me among Turner's sea pieces, on account of the noble gathering together of the great wave on the left,—the back of a breaker, just heaving itself up, and provoking itself into passion, before its leap and roar against the beach. But the enjoyment of these designs is much interfered with by their monotony : it is seriously to be regretted that in all but one the view is taken from the sea ; for the spectator is necessarily tired by the perpetual rush and sparkle of water, and ceases to be impressed by it. It would be felt, if this plate were seen alone, that there are few marine paintings in which the weight and heaping of the sea are given so faithfully.

For the rest it is perhaps more to be re-gretted that we are kept to our sea-level at

Sidmouth than at any other of the localities
illustrated.   What claim the pretty little
village has to be considered as a port of
England, I know not; but if it was to be
so ranked, a far more interesting study of it
might have been made from the heights above
the town, whence the ranges of dark-red
sandstone cliffs stretching to the south-west
are singularly bold and varied.   The detached
fragment of sandstone which forms the principal object in Turner's view has long ago
fallen, and even while it stood could hardly
have been worth the honour of so careful
illustration.

WHITBY

X.—WHITBY

As an expression of the general spirit of English coast scenery, this plate must be considered the principal one of the series. Like all the rest, it is a little too grand for its subject; but the exaggerations of space and size are more allowable here than in the others, as partly necessary to convey the feeling of danger conquered by activity and commerce, which characterises all our northerly Eastern coast. There are cliffs more terrible, and winds more wild, on other shores; but nowhere else do so many white sails lean against the bleak wind, and glide across the cliff shadows. Nor do I know many other memorials of monastic life so striking as the abbey on that dark headland. We are apt in our journeys through lowland

England, to watch with some secret contempt the general pleasantness of the vales in which our abbeys were founded, without taking any pains to inquire into the particular circumstances which directed or compelled the choice of the monks, and without reflecting that, if the choice were a selfish one, the selfishness is that of the English lowlander turning monk, not that of monachism; since, if we examine the sites of the Swiss monasteries and convents, we shall always find the snow lying round them in July; and it must have been cold meditating in these cloisters of St. Hilda's when the winter wind set from the east. It is long since I was at Whitby, and I am not sure whether Turner is right in giving so monotonous and severe verticality to the cliff above which the abbey stands; but I believe it must have some steep places about it, since the tradition which, in nearly all parts of the island where fossil ammonites are found, is sure to be current respecting them, takes quite an original form at Whitby, owing to the steepness of this rock. In general, the saint of the locality has simply turned all the serpents to stone ; but at Whitby, St. Hilda drove them over the cliff, and the serpents, before being petrified, had all their heads broken off by the fall !

DEAL

XI.—DEAL

I HAVE had occasion,* elsewhere, to consider
at some length, the peculiar love of the
English for neatness and minuteness : but I
have only considered, without accounting for,
or coming to any conclusion about it ; and,
the more I think of it, the more it puzzles
me to understand what there can be in our
great national mind which delights to such
an extent in brass plates, red bricks, square
kerbstones, and fresh green paint, all on the
tiniest possible scale.   The other day I was
dining in a respectable English " Inn and
Posting-house," not ten miles from London,
and, measuring the room after dinner, I
found it exactly twice and a quarter the
height of my umbrella.   It was a highly
comfortable room, and associated, in the
proper English manner, with outdoor sports

* *Modern Painters*, vol. iv. chap. I.

and pastimes, by a portrait of Jack Hall, fisherman of Eton, and of Mr. C. Davis on his favourite mare ; but why all this hunting and fishing enthusiasm should like to reduce itself, at home, into twice and a quarter the height of an umbrella, I could not in any wise then, nor have I at any other time been able to ascertain.

Perhaps the town of Deal involves as much of this question in its aspect and repu- tation, as any other place in Her Majesty's dominions : or at least it seemed so to me, coming to it as I did, after having been accustomed to the boat-life at Venice, where the heavy craft, massy in build and massy in sail, and disorderly in aquatic economy, reach with their mast-vanes only to the first stories of the huge marble palaces they anchor among.   It was very strange to me, after this, knowing that whatever was brave and strong in the English sailor was concentrated in our Deal boatmen, to walk along that trim strip of conventional beach, which the sea itself seems to wash in a methodical manner, one shingle-step at a time ; and by its thin toy-like boats, each with its head to sea, at regular intervals, looking like things that one would give a clever boy to play with in a pond, when first he got past petticoats; and the row of lath cots behind, all tidiness and telegraph, looking as if the whole business of the human race on earth was to know what o'clock it was, and when it would be high water,—only some slight weakness in favour of grog being indicated here and there by a hospitable - looking open door, a gay bow-window, and a sign intimating that it is a sailor's duty to be not only accurate, but "jolly."

Turner was always fond of this neat, courageous, benevolent, merry, methodical Deal. He painted it very early, in the Southern Coast series, insisting on one of the tavern windows as the principal subject, with a flash of forked lightning streaming beyond it out at sea like a narrow flag. He has the same association in his mind in the present plate; disorder and distress among the ships on the left, with the boat going out to help them; and the precision of the little town stretching in sunshine along the beach.

SCARBOROUGH

XII.—SCARBOROUGH

I HAVE put this plate last in the series, think-
ing that the reader will be glad to rest in
its morning quietness, after so much tossing
among the troubled foam. I said in the course
of the introduction, that nothing is so perfectly
calm as Turner's calmness; and I know very
few better examples of this calmness than the
plate before us, uniting, as it does, the glittering of the morning clouds, and trembling of
the sea, with an infinitude of peace in both.
There are one or two points of interest in
the artifices by which the intense effect of
calm is produced. Much is owing, in the
first place, to the amount of absolute gloom
obtained by the local blackness of the boats
on the beach; like a piece of the midnight
left unbroken by the dawn. But more is
owing to the treatment of the distant harbour mouth. In general, throughout nature,

Reflection and Repetition are *peaceful* things;
that is to say, the image of any object, seen
in calm water, gives us an impression of quietness, not merely because we know the water
must be quiet in order to be reflective; but
because the fact of the repetition of this form
is lulling to us in its monotony, and associated
more or less with an idea of quiet succession,
or reproduction, in events or things throughout
nature :—that one day should be like another
day, one town the image of another town, or
one history the repetition of another history,
being more or less results of quietness, while
dissimilarity and non-succession are also, more
or less, results of interference and disquietude.
And thus, though an echo actually increases
the quantity of sound heard, its repetition of
the notes or syllables of sound, gives an idea
of calmness attainable in no other way ; hence
the feeling of calm given to a landscape by the
notes of the cuckoo.   Understanding this, observe the anxious *doubling* of every object by a
visible echo or shadow throughout this picture.
The grandest feature of jt is the steep distant
cliff; and therefore the dualism is more marked
here than elsewhere ; the two promontories

or cliffs, and two piers below them, being arranged so that the one looks almost like the shadow of the other, cast irregularly on mist. In all probability, the more distant pier would in reality, unless it is very greatly higher than the near one, have been lowered by perspective so as not to continue in the same longitudinal line at the top,—but Turner will not have it so; he reduces them to exactly the same level, so that the one looks like the phantom of the other; and so of the cliffs above.

Then observe, each pier has, just below the head of it, in a vertical line, another important object, one a buoy, and the other a stooping figure. These carry on the double group in the calmest way, obeying the general law of vertical reflection, and throw down two long shadows on the near beach. The intenseness of the parallelism would catch the eye in a moment, but for the lighthouse, which breaks the group and prevents the artifice from being too open. Next come the two heads of boats, with their two bowsprits, and the two masts of the one farthest off, all monotonously double, but for the diagonal mast of the nearer one, which again hides the artifice. Next, put your finger over the white central figure, and follow the minor incidents round the beach; first, under the lighthouse, a stick, with its echo below a little to the right; above, a black stone, and its echo to the right; under the white figure, another stick, with its echo to the left; then a starfish,* and a white spot its echo to the left; then a dog, and a basket to double its light; above, a fisherman, and his wife for an echo; above them, two lines of curved shingle; above them, two small black figures; above them, two unfinished ships, and two forked masts; above the forked masts, a house with two gables, and its echo exactly over it in two gables more; next to the right, two fishing-boats with sails down; farther on, two fishing-boats with sails up, each with its little white reflection below; then two larger ships, which, lest his trick should be found out, Turner puts a dim third between; then below,

---

* I have mentioned elsewhere that Turner was fond of this subject of Scarborough, and that there are four drawings of it by him, if not more, under different effects, having this much common to the four, that there is always a starfish on the beach.

two fat colliers, leaning away from each other, and two thinner colliers, leaning towards each other ; and now at last, having doubled everything all round the beach, he gives one strong single stroke to gather all together, places his solitary central white figure, and the Calm is complete.

It is also to be noticed, that not only the definite repetition has a power of expressing serenity, but even the slight sense of *confusion* induced by the continual doubling is useful ; it makes us feel not well awake, drowsy, and as if we were out too early, and had to rub our eyes yet a little, before we could make out whether there were really two boats or one.

I do not mean that every means which we may possibly take to enable ourselves to see things double, will be always the most likely to ensure the ultimate tranquillity of the scene, neither that any such artifice as this would be of avail, without the tender and loving drawing of the things themselves, and of the light that bathes them ; nevertheless the highest art is full of these little cunnings, and it is only by the help of them that it can succeed in at all equalling the force of the natural impression.

One great monotony, that of the successive sigh and vanishing of the slow waves upon the sand, no art can render to us. Perhaps the silence of early light, even on the "field dew consecrate" of the grass itself, is not so tender as the lisp of the sweet belled lips of the clear waves in their following patience. We will leave the shore as their silver fringes fade upon it, desiring thus, as far as may be, to remember the sea. We have regarded it perhaps too often as an enemy to be subdued; let us, at least this once, accept from it, and from the soft light beyond the cliffs above, the image of the state of a perfect Human Spirit,—

> "The memory, like a cloudless air,
> The conscience, like a sea at rest."

THE END.

Printed by BALLANTYNE, HANSON & CO.
Edinburgh and London

www.ingramcontent.com/pod-product-compliance
Lightning Source LLC
Chambersburg PA
CBHW030845270326
41928CB00007B/1221